CATTLE IN THE COTTON FIELDS

CATTLE IN THE COTTON FIELDS

A History of Cattle Raising in Alabama

BROOKS BLEVINS

THE UNIVERSITY OF ALABAMA PRESS
Tuscaloosa and London

Copyright © 1998
The University of Alabama Press
Tuscaloosa, Alabama 35487-0380
All rights reserved
Manufactured in the United States of America

Portions of Chapter 2 have been published in the October 1998 issue
of *The Alabama Review* © 1998 The University of Alabama Press.

∞

The paper on which this book is printed meets the minimum requirements of
American National Standard for Information Science–Permanence of Paper for Printed
Library Materials, ANSI Z39.48-1984.

Library of Congress Cataloging-in-Publication Data

Blevins, Brooks, 1969–
Cattle in the cotton fields : a history of cattle raising in
Alabama / Brooks Blevins.
p. cm.
Includes bibliographical references (p.   ) and index.
1. Cattle—Alabama—History. I. Title.
SF196.U5 B58 1998
636.2′009761—dc21
98-19778

British Library Cataloguing-in-Publication Data available

ISBN 0-8173-0940-3

To the Memory of
BRYAN AND ALVERDA BLEVINS

CONTENTS

Preface	ix
1. The Melding of Traditions	1
2. Piney Woods and Plantations	13
3. Agricultural Progressivism and the South	43
4. The Midwestern Model Meets the South	76
5. Cattle in the Cotton Fields	114
6. New Farmers in the New South	143
Appendix	167
Notes	175
Bibliography	197
Index	207

MAPS AND TABLES

Maps

Alabama Counties	xiii
Physiographic Regions of Alabama	xiv

Tables

Alabama Cattle and Human Populations, 1840–1990	167
Effects of the Boll Weevil in the Black Belt, 1910–1920	168
Cattle and Cotton Prices, Selected Years	169
Presidents of the Alabama Cattlemen's Association	170
Cattle Numbers in Selected Counties, 1959–1992	172
Cattle-Raising Statistics, Selected Counties, 1992	173

PREFACE

In 1994 cattlemen's groups in Alabama and across the nation celebrated the 500th anniversary of the arrival of cattle in the Americas. Undoubtedly, cattle raisers of the fifteenth century, or piney woods herders of the early twentieth century for that matter, would find familiar little if anything in the modern cattle industry. Larger, meatier animals roam pastured fields enclosed by barbed-wire fences. Roundups and brands are as rare in Alabama today as purebred Herefords were in the nineteenth century. The forces of government and science have cooperated to control diseases and parasites, and American wealth and affluence have created an unimaginable demand for beef.

In Alabama the cattle raisers of the late twentieth century bear little resemblance to those of the early twentieth century, and much less to the native American and French herders of the eighteenth century. Open-range herding continued throughout the state until Reconstruction. Beginning in the plantation districts and spreading into the less fertile regions, planters and large farmers accomplished a steady encroachment upon the range and its herders. Simultaneously, federal and state governments laid the foundation for the development of scientific agriculture, research, and extension. The southern economic morass after the Civil War reinforced the dependency on cotton and the development of the system of tenancy and sharecropping. Only with the arrival of the boll weevil and Roosevelt's New Deal did King Cotton find itself threatened by other agricultural commodities.

Planters of the Black Belt, a long fertile stretch of western and central Alabama, because of the region's unique geological characteristics, first abandoned cotton in significant numbers upon the arrival of the boll weevil. In the three decades after World War I many Black Belt planters adopted a midwestern model of cattle raising consisting of purebred British breeds, improved and enclosed pastures, scientific breeding and feeding practices, and intimate cooperation between cattlemen, government agents, and business interests. These modern cattle raisers and their practices differed strikingly from those of the open-range herding tradition, which slowly disappeared before the closing of the state range in 1951. Nonetheless, these twentieth-century cattlemen inherited the practices and values of their planter forefathers, and their combination of midwestern methods with traditional southern labor practices reflected the continuity of the planter spirit.

In the past half century the Alabama cattle industry has witnessed tremendous growth, and revenue generated by the state's cattle raisers is surpassed only by that of the poultry industry among Alabama's agricultural commodities. The cattle-raising industry has played an integral role in agricultural transformation and social change and has reflected the important position of such government agencies as the cooperative extension service and agricultural experiment station. Cattle raising has also undergone demographic changes. Since the 1960s other regions of the state, most notably the Appalachian counties, have equaled and often surpassed the once dominant Black Belt in cattle production. A key part of this phenomenon has been the increasing importance of part-time farmers in the cattle-raising business. The absence of federal price controls and subsidies for cattle raisers has made the business a precarious pursuit, one especially suited

PREFACE

to individuals obtaining their primary sources of income elsewhere. As a result commercial cattle raising, dominated by plantation-belt planters in the mid–twentieth century, has in the past three decades increasingly regained the egalitarian characteristics of the antebellum era and has become the state's most popularly practiced agricultural pursuit.

The work that follows grew out of a research project first started in the summer of 1994. In August of that year I began the task of collecting historical information on the Alabama cattle industry and organizing this information for the Alabama Cattlemen's Association's historical museum. In the process the story of the Alabama cattle industry began to unfold, offering both familiar themes and unexpected developments. This study is agricultural history set within the context of southern history, and therefore I attempt to accomplish some balance and interconnectedness between the popular themes associated with each field. This is not an economic history, though in places statistics and percentages abound. Because this is the first lengthy study of the development of the cattle industry in a southern state—and possibly in any state—I present the history of Alabama cattle raising in a chronological narrative and interject historiographical interpretations and themes when pertinent. Perhaps this method will best provide an introduction to an overlooked and underworked topic in agriculture and in the development of the South.

Over the course of this project I have received invaluable aid and advice from numerous people. For first alerting me to the project in the spring of 1994, I owe a debt of gratitude to Marty Olliff. For their useful critiques of chapters or sections of the book, I thank Tony Carey, Larry Gerber, Steve Murray, Carol Ann Vaughn, and Gordon Harvey. Chuck Simon enthusiastically lent his vast knowledge of antebellum cattle herding in the early stages of my work. The Alabama Cattlemen's Association provided generous financial support for two years of research and writing, and Dr. Billy Powell, Cynthia Townley, and Meg Truman provided useful assistance and direction. Dwayne Cox, Beverly Powers, and the staff of the Auburn University Archives graciously assisted in my numerous and sometimes odd requests of their time and expertise, as did Norwood Kerr and the staff at the Alabama Department of Archives and History in Montgomery. Melba Fulbright, Tom Wilkins, and the faculty at Mount Pleasant (Arkansas) School graciously provided me with word-processing capabilities and the knowledge to use them. Finally, I would like to thank the two people without whom this book would not have been written: Wayne Flynt, whose careful readings and sincere direction helped me through the rocky spots, and my wife Sharon, whose unconditional and constant moral support, in spite of her disinterest in the subject, always kept me going.

Alabama Counties

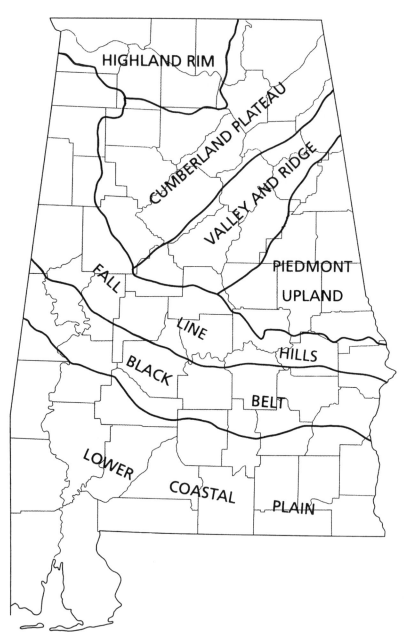

Physiographic Regions of Alabama

CATTLE IN THE COTTON FIELDS

Chapter 1

The Melding of Traditions

Although historians do not know when cattle first stepped onto the soil of present-day Alabama, several clues allow us to speculate on the bovine's arrival. Over a century and a half before the French settled Dauphin Island and Mobile Bay, Spanish explorers and missionaries began bringing cattle to the North American mainland.

One of the earliest means of livestock introduction was the Spanish exploration party. Ponce de León brought cattle and swine on his second trip to Florida in 1521. The livestock were meant to sustain the explorers, and most likely few survived to establish a permanent herd. Though the surviving records of Hernando de Soto's expedition through the southeastern United States in 1539 list hogs and horses but no cattle, subsequent expeditions did travel with herds of cattle. In 1540, Don Diego Maldonado

2
THE MELDING OF TRADITIONS

transported a number of cattle to the Pensacola area, as did Don Tristan de Luna in 1559.[1]

By the time of de Luna's expedition the Spanish had realized the potential for large-scale cattle raising in Florida. In late 1559 Luis de Velasco attempted to organize a drive of 5,000 head of cattle from northern Mexico to de Luna's province of Pensacola Bay. The demands of mere subsistence in the untamed land soon dashed such dreams, but de Luna continued to request the animals by sea if not by land. In the spring of 1560 de Luna referred to horses and cattle as the colonists' "two most essential" needs. Though Velasco continued to promise to send cattle from Mexico or Havana to de Luna's struggling party, the occupants of Pensacola Bay "were reduced to eating the hides of cattle which they had brought from New Spain and all the Horses that they had brought."[2]

In spite of the difficulties experienced by early explorers and settlers, their outposts provided convenient centers from which to spread their culture and agriculture into the surrounding areas occupied by native Americans. In the late sixteenth century Spanish missionaries traveled up rivers into Alabama and Georgia to establish mission villages; these missionaries took livestock with them and likely introduced the practice of cattle herding to natives of the region. These native Americans quickly found the white man's animals good for some uses. In the late 1500s Spanish colonists found cowhides and beef among their American neighbors.[3]

Despite the existence of a small number of tame cattle among the Spanish colonists and a few native American groups and an unknown number of feral cattle in the southern piney woods, no substantial herding took place in Spanish Florida until the mid–seventeenth century. The last two decades of the seventeenth century witnessed a cattle boom in Florida, a boom that likely exercised tremendous influence on the living habits of neighboring native American tribes as well as tribes in southern Alabama and Georgia. According to geographer Terry G. Jordan, cattle ranching in Florida closely resembled the Spanish style of the West Indies. This method of ranching, descended from the Andalusian and Antillean regions of Spain, was taken on in a modified form by the Seminoles and Creeks in the first half of the eighteenth century.[4]

Though feral and tame cattle undoubtedly roamed the Alabama countryside by 1701, the French made the first documented introduction of cattle onto Alabama soil in that year. A French Canadian expedition, headed by brothers Pierre Le Moyne d'Iberville and Jean-Baptiste Le Moyne d'Bienville, had begun colonization efforts on Biloxi Bay and Dauphin Island in 1699. Within the next two years Iberville transported more than twenty Spanish longhorns with several hogs and horses from the West Indian port of Santo Domingo to the Mobile area, probably Dauphin Island. By 1709 there were over 100 head of cattle in French Lower Louisiana, which included the Mobile area. One early settler, Jean-Baptiste Boudreau de Graveline, recognized the possibilities for cattle raising on Dauphin Island. Graveline stopped in Havana to purchase cattle on his first two trips to the young colony before finally settling on the island and building a good herd. By 1713 some 300 cattle ranged on Dauphin Island and at Fort St. Louis (Mobile) on the mainland.[5]

Once firmly established on Mobile Bay, many French settlers turned their attentions toward establishing a profitable cattle industry. In 1713 two men traveled from Mobile to Vera Cruz, Mexico, for the purpose of trading French merchandise for cattle and horses. Unable to secure the desired stock from the Spanish territories, the French looked eastward to the British in the Carolinas. In October 1719, a council of commerce assembled on Dauphin Island "decided to obtain [cattle] by land from Carolina and to write to the governor of that country [to ask] whether he will have some furnished at what price and whether in payment he wishes deerskins or letters on the Western Company at Paris." The council feared the "colony will never be well established without this essential assistance." Earlier that month governor Bienville had ordered the construction of a butcher shop on Dauphin Island to provide fresh beef for inhabitants of the island and the mainland.[6] The French at Mobile eventually obtained livestock from various places, including Mexico, Texas, Florida, Cuba, and even the French in the Illinois country. By the 1720s, according to Terry G. Jordan, Mobile enjoyed an enduring cattle industry. One historian even claims "the business of herding was becoming almost as important under the French as among the Spaniards further South."[7]

Like the Spanish before them, the French helped spread the herding culture to native tribes of Alabama. French missionaries and traders traveled up the waterways of the Alabama, Tombigbee, and other rivers, spreading their merchandise and agricultural methods among the Choctaws,

Creeks, and other peoples. The eighteenth century witnessed the tremendous rise of native American cattle raising. Though the Creeks and Choctaws are most often noted as cattle raisers, a French cartographer found large stocks of cattle among the Cherokees in 1712. The Cherokees, located hundreds of miles north and east of Mobile, most likely obtained their cattle and methods of herding from British settlers in the Carolina back country.[8]

The South Carolina cattle industry blossomed in the last quarter of the seventeenth century west and southwest of Charleston, and by 1760 cowpens could be found as far into the back country as the fall line. Carolina cattle culture, though influenced by the Jamaican longhorn herding system, relied primarily on British stock and British methods, such as tending cattle on foot and using salt to control and lure herds. By the time Great Britain obtained all of Alabama in 1763, opening the way for settlement by these Carolina cattlemen, the French and their native American neighbors had established a significant cattle industry.[9]

British and Anglo-American herders trickled into Alabama after 1763 and further altered the conglomerate of herding practices. During their twenty-year occupation of Mobile, British colonists increased their livestock holdings and built Mobile into a cow town. In 1767 the governor of the colony declared it illegal to drive the cattle from the vicinity west of Mobile all the way to Louisiana for sale. A few cattlemen in the area owned herds of up to 1,000 head of cattle. A 1766 census of seventeen plantations on the east side of Mobile Bay counted 2,280 head of cattle and only 124 people. One decade later Scottish settler Patrick Strachan owned 600 head of cattle on his 11,000-acre Tensaw River plantation, and Elias Durnford claimed a herd of 1,400 head east of Mobile Bay. In the 1760s some Mobile cattlemen supplied Pensacola with beef, though they required a fort for protection from attacks by native Americans. Neighboring native Americans could also pose a threat to the British colonists' livestock. Daniel Hickey, plantation overseer for British West Florida Lieutenant Governor Montfort Browne, reported the Creeks' and Alabamas' habit of rowing boats to Dauphin Island and killing cattle for food and hides. He counted the loss at 114 head, only a small percentage of the island's cattle.[10]

With the growing tide of Anglo-American and European settlers entering Alabama in the late 1700s, the newcomers and natives found

THE MELDING OF TRADITIONS

themselves inextricably bound together in their struggles over land and traditional practices. At the center of many confrontations were issues of grazing rights and stolen stock.

A 1771 congress between British authorities in West Florida and representatives of the Creek nation reflected the difficulties that arose from the arrival of cattle herders from the east. British authorities assembled the group in an attempt to halt the Creeks' practice of killing the white men's straggling cows. John Stuart, head of the British delegation, attempted to reassure his Creek visitors: "Surely my Brethren no damage or prejudice can arise to your nation by a few Cows straggling into your Woods and eating a Little Grass." Chief Emistisiugo quickly asserted that he was not referring to "a few Cows" but several herds of cattle: "Besides Mr. Galphin, who was the first that drove Cattle thro' our Nation, there are many others driving Cattle and Settling Cowpens on our Land without our Consent." After naming eight cattlemen who had settled within Creek territory, Emistisiugo accused Stuart of breaking a 1765 agreement when he had his cattle driven through Creek lands on the way to his plantation east of Mobile Bay. Furthermore, the chief complained of a herder whom the Cherokees had allowed to settle along the upper forks of the Coosa River, from which location he drove cattle into Creek territory.[11]

When Creek agent Benjamin Hawkins came to Alabama in the 1790s, he found native-white struggles over livestock continuing. In some cases he reported the tables had turned—the Creeks were often the large herders and white settlers the thieves. An important change had occurred in many Creek villages and among the peoples of other nations. Although Creeks and other native Americans had been introduced to cattle generations earlier, the arrival of Anglo-American herders on their lands and the declining numbers of wild game animals hastened the natives' reliance on cattle herding. Within two generations after the arrival of Anglo-American pioneers, Alabama's native tribes had replaced the traditional hunting and gathering culture with a herding culture. According to Michael F. Doran, "Both animals and the herding tradition were among the first Anglo-American traits accepted."[12]

Hawkins observed this native American herding phenomenon on his trek through Creek territory in 1798 and 1799. Hawkins noted the excellent range along the lower Alabama River and between the Flint and

Chattahoochee rivers. He also found an abundance of switch cane and creek moss in the area between the Tallapoosa and Coosa rivers and the Chattahoochee. Hawkins reported cattle—often in herds of more than 100 head—as integral holdings of almost every village he visited. Noting early in his journey that the Creeks had recently begun settling in villages because of its convenience for stock raising, Hawkins found several villages that excelled at cattle raising. At the conjunction of the Coosa and Tallapoosa rivers Hawkins found a Creek town with more cattle than perhaps any other town he visited. At Ocfuskee on the upper Tallapoosa he found the largest and finest cattle in the nation. Near present-day Tallassee Chief Toolkaubatche Haujo owned around 500 head and "seldom kill[ed] less than two large beeves a fortnight, for his friends and acquaintances." Though Hawkins found the most and best cattle along the Tallapoosa, he also witnessed fine stocks among the natives on the Coosa and Flint rivers, as well as on the upper Alabama. The agent concluded that the Creeks of southeastern Alabama also held livestock in high regard and noted that they marketed their cattle, hogs, and poultry at Pensacola.[13]

Mixed-race Anglo-Creeks usually led the way in cattle herding. It was often their white fathers who had settled among the natives with their livestock. Hawkins mentioned many such herders in his travel account. Peter McQueen possessed a number of cattle, hogs, and horses near Tallassee. Near present-day Wetumpka Anglo-Creek Sam Macnack owned 180 calves alone in 1799. In one village on the Tallapoosa Hawkins met four mixed-race herders with large stocks of cattle, as well as Robert Grierson, a Scot married to a Creek woman; Grierson possessed 300 head of cattle and thirty horses. Daniel McGillivray, son of a Scottish merchant and a Creek woman, lived on a plantation on the lower Coosa. McGillivray owned a large herd of cattle and frequently sent factors to sell his cattle and deerskins at Pensacola. Other areas of the state had significant numbers of mixed-race herders as well. On the Little River in southwestern Alabama there lived several mixed-race families. Milly, a white woman whose British husband had died, married a native American, and the two raised cattle and horses. Another was William Gregory, who married a native American woman and maintained a large herd of cattle.[14]

Once settled into his job as Creek agent, Hawkins faced many of the problems British officials had wrestled with thirty years earlier. In an 1800 letter to Thomas Jefferson, Hawkins claimed the Creeks in his area had ready for market 1,000 head of beef cattle and 300 hogs, but that they had begun "to be attentive to the raiding of stock." In 1805, Hawkins informed the governor of Georgia that several Creeks had seen "white people stealing cattle repeatedly." And, harking back to observations of the disgruntled Chief Emistisiugo, Hawkins explained that most of the stock stolen by the squatters were stray cattle belonging to frontier property owners. One chief wrote to Hawkins in 1808 to complain of a Colonel Easley who had erected a cowpen on Creek land. Furthermore, decried the chief, "Our hunters are now in their woods hunting as usual for game, and they find cattle ranging in large droves on their hunting grounds, and salt logs where the intruders came to salt their cattle." Three months later two chiefs informed Hawkins that four white men had released over 100 head each on Creek land to mix with Creek cattle. The chiefs worried that "the white owners in gathering their stock will be as they have generally been, not exact in taking none but their own."[15]

Hostilities existed not just between whites and native Americans. The Creek Wars of 1813 and 1814 provided warring tribes with ample opportunity to steal and slaughter the cattle of neighboring whites as well as those of enemy tribes. In the summer of 1813, according to Hawkins, the "Fanatical Chiefs" gathered on the lower Tallapoosa and raided their neighbors' cattle herds. Less than one month later Hawkins informed Captain Carr that "the Fanatics have destroyed some of the largest stock of the Upper Creeks." After the war Hawkins found southeastern Alabama and southwestern Georgia "destitute of food . . . not a horse, hog or cow to be seen."[16]

While the Creeks fought with each other and with American forces, Anglo-American herders continued to migrate across the pine belt of Georgia into the southern Alabama piney woods and, in fewer numbers, into the Piedmont and Black Belt. The Great Valley brought livestock herders into northeastern Alabama. Highlanders began bringing their hogs and cattle into the mountains in the northern part of the state, and Georgians and Tennesseans found the Tennessee Valley inviting.

The period between the American Revolution and Alabama's statehood in 1819 witnessed a massive immigration of Anglo-American herders, small farmers, and planters. Before 1800 white settlements in Alabama had been confined to the southwestern corner, with the exception of those persons living among the native Americans and a few scattered, sturdy pioneers in the interior. Southwestern Alabama had long been the site of a thriving cattle business. By the late 1790s along the lower Tombigbee cattle outnumbered people five to one, and Mobile's chief exports had for some time included "fine cattle . . . and salted wild beef." The waves of white settlers, usually led by herders and drovers, took livestock with them into areas previously uninhabited by whites.[17]

Because of the scarcity of census records and written accounts from inhabitants, the journals of explorers and travelers of the late eighteenth and early nineteenth centuries offer the historian an invaluable look at the people and customs of the Old Southwest. As a result of the importance of the cattle industry in the region, chroniclers in Alabama between the 1770s and the 1830s made several references to livestock and cattle raisers.

On a journey through western Florida and southwestern Alabama in 1770 and 1771, British traveler Edward Mease visited a small French village near Mobile where the inhabitants raised cattle that looked "Poor like the Country." Like others before and after, Mease observed a fine range for cattle on Dauphin Island. North of Mobile his party followed a "Cattle Path" and passed by "two or three Cane Breaks where there was the Appearance of much Cattle having been drove."[18]

Naturalist Bernard Romans offered a more critical and analytical estimation of the region in his book *A Concise Natural History of East and West Florida*. Romans discovered "incomparable pasture" along the Perdido River and its tributaries and estimated the number of cattle and horses between the Perdido and Tensaw rivers at 10,000 head. Romans clearly disliked the southerners' neglect of dairying: "The manner in which cattle are now kept in the southern colonies is unprofitable; twelve or sixteen cattle might be with a little attention made to yield more profit in dairy, than flocks of three or four hundred do now, with all the labour and time at present bestowed on them." The naturalist did, nevertheless, realize the potential for other cattle uses: "When we consider the vast increases of stock here, and the ease of maintaining a stock of a thousand or fifteen hundred cattle . . . it must be evident, that tanning of leather will be a great business here."[19]

William Bartram, in his extensive travels throughout the Southeast, reported several revealing stories. In 1776 Bartram observed a large dairy and beef operation in eastern Georgia. The owner maintained some 1,500 head of cattle, forty of which his slaves milked daily and maintained within movable cowpens. The herder used the milk only for his and his slaves' families and annually sold a number of his stock for cash. Such operations, though perhaps on a smaller scale, could be found throughout the Southeast in the era. Bartram found "exuberant pasture for cattle" around Pensacola and met traders in southern Alabama who carried whips made of cowhide.[20]

John Pope journeyed through the Creek territory in the early 1790s. In 1791 he visited Alexander McGillivray on his Coosa River plantation. Pope observed that two or three white cowhands supervised McGillivray's range. Baiting their pens with salt, the hands would "now and then collect [the cattle] together in Order to brand, mark, etc." Pope also noted the existence of a Creek man, Bully, who owned over 500 head of cattle and 250 horses.[21]

Travel literature from the early nineteenth century provides other quick glances at a now more familiar land. In 1807 Englishman John Melish observed that in the deep South "the cattle graze in the fields all winter, a circumstance highly favorable to the husbandmen." Another British traveler, James Stuart, deemed the tall, coarse prairie grass between Columbus, Georgia, and Montgomery "succulent and rich food for cattle," but the amazed visitor found few cattle on his trek. A few years later Harriet Martineau enjoyed a large breakfast of beefsteaks and plenty else on her stop in Montgomery, a pleasant meal not often repeated on her journey. Nevertheless, complained the British woman, a prairie weed eaten by cows in the summer "makes the milk so disagreeable, that cream, half-an-inch thick, is thrown to the pigs."[22]

Outside of the native American territories, Alabama cattle raising in the late 1700s and early 1800s became increasingly the domain of Anglo-American settlers. Though cattle herders wandered into all regions of the state, not all areas participated equally in the growing cattle industry. Cotton planters rushed to occupy the fertile soils of the Black Belt and Tennessee Valley regions after the federal government placed Mississippi Territory land on sale in 1807. Between 1810 and 1820 the combined population of Mississippi and Alabama catapulted from 40,000 to 220,000, largely powered by the cotton boom. The Piedmont as well gradually found itself overrun by small cotton growers and yeoman farmers. In that region, as in the mountains to the north, hogs came to dominate livestock raising.[23]

The southern piney woods area provided an ideal habitat for cattle raisers. Warm winters allowed for year-round grazing with little attention to stock. Longleaf pine forests presented little allure for the planter but offered saw grass and other pastures for cattlemen. Cane brakes also abounded, as did water sources. Consequently, southern Alabama, which already boasted a cattle industry several generations old by the time Anglo-American pioneers arrived, became Alabama's leading cattle region. An 1804 census of Washington County, Mississippi Territory, which included most of southern Alabama and southeastern Mississippi, revealed several large herders in southwestern Alabama. In the Tensaw district about thirty families owned herds of at least 100 head of cattle, and a cattleman named John Randon owned 1,000 head. Cattle herds in this area were dominated for many years by Spanish longhorns, now referred to as cracker cattle. These rugged animals were well suited for the open-range, Spanish-British herding style of the early cattlemen.[24]

Families owning large herds of cattle frequently maintained "cowpens." Though the term would eventually come to denote a specific, enclosed spot of land on a farm or plantation, in the colonial era a cowpen generally described the headquarters of a ranch. Though descriptions of eighteenth-century cowpens are scarce, a British officer left a colorful account of a cowpen in the colonial back country during the 1750s:

> From the Heart of the Settlements we are now got into the Cow-pens; the Keepers of these are very extraordinary Kind of Fellows, they drive up their Herds on Horseback, and they had need do so, for their Cattle are near as wild as Deer; a Cow-pen generally consists of a very large Cottage or House in the Woods, with about four-score or one hundred Acres, inclosed with high Rails and divided; a small Inclosure they keep for Corn, for the family, the rest is the Pasture in which they keep their calves . . . they may perhaps have a Stock of four or five hundred to a thousand Head of Cattle belonging to a Cow-pen, these run as they please in the great Woods, where there are no Inclosures to stop them. . . . Every Year in September and October they drive up the Market Steers, that are fat

· II ·
THE MELDING OF TRADITIONS

Native or "cracker" cow and calf in the piney woods of southern Alabama. (Courtesy of Alabama Cattlemen's Association)

and of a proper Age, and kill them; they reckon that out of 100 Head of Cattle they can kill about 10 or 12 steers, and four or five Cows a year; so they reckon that out of 100 Head of Cattle brings about £40 Sterling per Year. . . . The Cow-Pen Men are hardy People, are almost continually on Horseback, being obliged to know the Haunts of their Cattle.[25]

Harry Toulmin described cattle raising and cowpens in the Mobile area in an 1810 *American Register* series. Toulmin observed open pine woods "affording good range for cattle," and reported that the pine range meant no hay, corn, or salt was required for livestock. Many owners possessed herds of more than 500 head, which they drove into the forests away from settlements in the fall and collected at cowpens in the spring. Herders generally sold their beef to Spanish buyers in Mobile or Pensacola at the going price of three to three and a half cents per pound. Toulmin also noted that the harsh effects of heat and flies rendered milk production inadequate and caused cows to calve only once every two years.[26]

Further proof of the significance of the southern Alabama cattle industry can be found in the contents of claims on the federal government made by victims of the War of 1812. In seventy-six affidavits filed by citizens of the lower Tombigbee and Alabama river areas, losses of cattle amounted to 5,447 head worth an average of seven dollars each. The value of cattle lost accounted for almost one-third of the total loss recorded and more than twice as much as any other item. Historian John D. W. Guice concludes that the picture drawn from these affidavits "confirms the existence of a significant cattle industry in the area." Several claimants reported the loss of hundreds of head of cattle, including Charles Conway of Mobile County (750) and Moses Stedham of Baldwin County (500). Others included Don Miguel Eslava, a Spaniard who had owned the largest herd east of Mobile Bay before the war, and the Trouillets, free blacks who lost over 100 head of cattle. Guice also notes that the claims reveal an active trade in barreled beef, hides, leather, horns, and tallow, as well as the practice of tending stock on horseback in the region.[27]

On the eve of Alabama's statehood in 1819 cattle played an integral role in the territory's agriculture and economy. A host of cultures and nationalities had contributed to herding systems throughout the state. Spanish, French, British, native American, and African influences shaped an age-old practice of husbandry to a new land. Pioneer yeomen from the mountains to the swampy lowlands depended on cattle for labor, food, clothing, and more. Cattle-raising communities in the southern piney woods and scattered herders throughout the state relied on their stock for income and sustenance. But Alabama, like the rest of the young South, was experiencing an agricultural revolution. This revolution would witness the breakneck expansion of cotton planting into all regions of the state, a development that would be reversed only a century and more later. Cattle herding continued to grow and become identified by region, but it would do so in obscurity as the South's eyes turned to the white fiber that would make her leaders rich.

Chapter 2

Piney Woods and Plantations

The history of agriculture in antebellum Alabama is generally portrayed as being synonymous with the history of cotton planting. Though cotton cultivation obviously dominated the attention of more and more agriculturists as the first half of the nineteenth century progressed, livestock raising continued to play an integral role in Alabama agriculture. The shape and significance of livestock raising were affected by numerous factors: region, socioeconomic status, and location in time. Alabama's cattle raisers, for instance, differed in many ways from region to region and from farm to farm over the half century in question.

A few historians have considered the importance of Alabama's and the South's antebellum cattle herders. Frank L. Owsley, John D. W. Guice, Terry G. Jordan, and John Solomon Otto view antebellum cattle raising as a dynamic process influenced by geography, settlement, and population growth. These historians accept a modified version of Frederick Jackson Turner's frontier herder thesis, though they build on Owsley's recognition of a permanent herder class in the southern backwoods, or inner frontiers. In so doing, these historians generally conclude that livestock raisers occupied frontier areas throughout the state in the late eighteenth and early nineteenth centuries but began to find their ranges significantly depleted by cotton farmers and planters as settlers flooded Alabama. By 1840, Owsley observes, the herdsmen had been driven from the better farming lands by growers of cotton.[1]

Other historians, such as Grady McWhiney, Forrest McDonald, and Eugene D. Genovese, tend to see cattle raisers and raising as static elements of antebellum society. McWhiney and McDonald find the southern herder in significant numbers throughout the regions of the South up to the Civil War. McWhiney attempts to refute Owsley's observation of a declining livestock herder culture and industry in antebellum Alabama. He stresses the continuity of the herdsman throughout the antebellum era and credits the Civil War with the devastation of the state's herding culture and industry.[2]

Historical evidence for Alabama appears to weigh in favor of the first group. Alabama's antebellum cattle industry changed continually throughout the period, growing in some areas and retreating in others. Owsley's observation that the encroachment of cotton growers pushed livestock raisers into the "inner frontiers" remains a plausible and valuable element of any thesis concerning the subject. In addition to the expansion of cotton cultivation, the shifting of frontiers and the growth of the Texas cattle kingdom threatened the future of open-range herding in Alabama and rendered traditional cattle raising a weakened and declining practice by the start of the Civil War.

McDonald and McWhiney identify three groups of livestock raisers in the antebellum South: planters, yeoman farmers, and herders or drovers. These three groups raised cattle in all sections of the state before the Civil War. Yeomen around the state usually maintained at least one milk cow for family use. Some small farmers also turned out a few head of cattle to graze on the open range. These cattle were used either for beef or as an occasional cash sale to a traveling drover. Planters generally maintained herds of cattle for plantation uses. As we shall see, many planters increased the quality and number of their herds in the late antebellum period.[3]

Much recent historical literature follows the trail of the antebellum southern herder, a pre–Civil War cattleman. Though the herder could be found in the different regions of Alabama, his domain was unquestionably the upland piney woods stretching across the southern part of the state. The earliest days of the nineteenth century likely saw significant numbers of cattle herders in the mountain valleys, the Tennessee Valley, the Piedmont, and the Black Belt. The steady encroachment of cotton-planting settlers in the fifty years before the Civil War pushed livestock raisers into Owsley's "inner frontiers." Hogs became the primary products of hill and mountain growers, while cattlemen generally found the piney woods more to their liking.

Perhaps it is a bit misleading, however, to speak of the herders as being driven deep into the barren piney woods. For many cattlemen the piney woods offered an ideal environment for herding. The warm climate eliminated the need for winter shelter, and the tall pine forests with little undergrowth provided better than adequate range pasture. Furthermore, swamp and river-bottom cane brakes provided valuable winter forage. Tradition also played an important role in the piney woods cattle industry. Not only had French, British, and Spanish settlers become accustomed to the herding benefits of southwestern Alabama, but the Anglo-American herders entering southeastern Alabama around 1800 also carried on the herding culture formed over a century earlier in the Carolina and Georgia backcountry.

Regardless of the route taken to reach the less arable soils of the piney woods, the most substantial number of the antebellum herdsmen were to be found in this belt making up most of the southern one-third of Alabama. Of those who had originally settled the rich bottomlands or better portions of the uplands, many probably settled down to become planters or yeoman farmers themselves. But the life of a cotton planter or yeoman farmer proved unsatisfying for countless herders, who continued to move west just ahead of the cotton growers or retreated into the least accessible areas of the pine forests.

Thus we discover a dynamic and complex assortment of cattle raisers with changing characteristics based on place and time. Although the piney woods region was home to the greatest number of cattlemen prior to the

Civil War, their numbers and their holdings differed from the early nineteenth century to the 1850s. Cotton expansion did not slow during the antebellum era, and the plant eventually found its way into every county in the state. Cotton planters conquered even the river bottoms of the piney woods area, further limiting the range of Alabama's cattle kings.

Despite the steady growth of cotton cultivation, Alabama's native tribes continued to occupy large sections of the state even after statehood. The Creeks had lost much of their western territory in the Treaty of Fort Jackson that followed the 1814 defeat at Horseshoe Bend; however, they retained thousands of acres in eastern and southeastern Alabama. Hundreds of Anglo-American families squatted on this land after 1814, often creating problems between whites and natives. The Cherokees occupied much of northeastern Alabama. That mountain nation, observed traveler James Stuart in the 1820s, contained some 22,000 cattle and almost 50,000 hogs. In western Alabama and eastern Mississippi one group of 5,600 Choctaws owned 11,000 cattle and 22,000 hogs.[4]

Pressure from white settlers hungry for native American lands soon forced federal and state government actions. After Congress formalized government policy with the Indian Removal Act of 1830, Alabamians lost no time in securing native American lands. A string of three treaties—Dancing Rabbit Creek (1830), Cusseta (1832), and New Echota (1835)—in five years removed from their traditional lands the Choctaws, Creeks, and Cherokees, respectively. Consequently, the 1830s witnessed another wave of white settlement on sparsely occupied Alabama lands.[5]

Other regions of Alabama were well settled by 1830. Southwestern Alabama and the Tennessee Valley were early destinations for white settlers. After 1810, the cotton boom brought families to the fertile Black Belt from Georgia and the Carolinas. The mountain and Piedmont areas not claimed by the Creeks and Cherokees also began to experience significant settlement. And white herders from the Georgia pine belt settled the Alabama piney woods.

Though the census bureau began collecting agricultural statistics only in 1840, several sources inform us of the substantial piney woods cattle industry in the early nineteenth century. Historian Lewis C. Gray identifies an 8,000-square-mile upland pine belt in southern Alabama that was settled by small farmers and herders. Frank L. Owsley notes that livestock ranged the uncultivated piney woods of Alabama up to the Civil War. This historian also realizes the mountain soils offered better ranges than those of the piney woods, but fewer cattle roamed the hillsides and mountains because of the rugged terrain. Historian John D. W. Guice suggests that cattle raising was the first profitable and lasting agricultural industry of southern Alabama and Mississippi. Furthermore, argues Guice, in that piney woods area until 1860 "more free persons probably sustained themselves by herding than by any other commercial means."[6]

In 1820, an English traveler in the piney woods of southern Alabama and northern Florida revealed the herder's reliance on marketing cattle in Mobile or Pensacola. Adam Hodgson dined at the home of a cattleman who owned some 2,000 head of cattle but "had neither milk nor butter to give us to our coffee." Hodgson further lamented his inability "to procure either milk or butter where 18 or 20 cows are kept." One historian of Clarke County, writing in the 1870s, recalled a Dr. Earle who had settled on the Tombigbee in 1824 with a large herd of cattle brought from Georgia. On his arrival in Clarke County in 1834, Tennessean David Byrd discovered plentiful herds of cattle in the area. Benjamin Franklin Riley, when considering the grasslands and cane brakes of Conecuh County in the 1880s, stated that "beef in considerable quantities . . . has for a long time been furnished from these, and adjoining regions, to the markets of Pensacola and Mobile."[7]

Southeastern Alabama also had its share of cattle raisers. Among them was William Cawthon, a Universalist preacher, businessman, and cattle baron who settled near Dothan and built a number of pine-pole cowpens. In the early days of settlement on the former Creek lands of Russell County, piney woods pioneers "made no corn the first few years, and but little of anything else, except stock, which ran wild on the public domain." The Reverend F. L. Cherry, writing in the 1880s, recalled a section of land "about a hundred square miles or more between the Chewahla and Uchee creeks which, fifty years ago, would not number more than a dozen families, and they were mostly 'cowboys.' This section was known as piney-woods, of Russell County, and . . . was considered very poor, and profitably available only as a stock range." Cherry also remembered the establishment of the Mimms trail for driving stock to market—probably to Columbus, Georgia—and from one range to another. Fred S. Watson, in his county history, calls antebellum Coffee County an "ideal cattle raising county."

Watson also notes that the Wiregrass counties of Covington, Coffee, Geneva, Henry, and Houston were for generations "known contemptuously as the 'cow counties.' "[8]

Often cattle raisers ranged their stock in the piney woods as far north as the edge of the Black Belt. Writing from southern Dallas County in 1838, Philip Henry Gosse recorded the memories of an "enthusiastic old sun-dried backwoodsman" who talked of "feral cattle existing in some of the more inaccessible swamps between these parts and the Florida border." The old man had been "engaged in parties to hunt them, shooting the cows for beef."[9]

Census records from the antebellum era reflect the importance of cattle raising in the piney woods. In 1850, the ten leading cattlemen of Covington County owned an average of more than 250 head. In the same year Washington County's top ten herders owned an average of 438 head, and at least one dozen cattlemen in that county owned more than 200 head.[10]

In both counties the largest herders devoted their energies almost exclusively to livestock raising. In Covington County, John Barrow owned 362 head of cattle, 13 sheep, and 60 hogs in 1850; he raised no cotton on his 700-acre farm. Another Covington County herder, Allen Hart, owned no land but ranged 284 cattle, 12 sheep, and 250 hogs. One Washington County cattleman had already begun to raise cotton by 1850, though it remained secondary to livestock on his 3,700-acre plot. A man named Johnston produced six bales of cotton and raised 1,200 cattle, 30 sheep, and 40 hogs. More typical of the piney woods herder in Washington County were Peter Laker and John Dearmon. From his 240-acre farm, Laker ranged 417 head of cattle and 100 hogs. Dearmon owned only forty acres but maintained a herd of 250 cattle and 100 hogs.[11]

The most colorful and descriptive account of piney woods herders was recorded by J. F. H. Claiborne in "A Trip through the Piney Woods." Though Claiborne's accounts illustrate his journey through southern Mississippi, they can be viewed as representative of the entire southern piney woods herding culture. Traveling in the 1840s in southeastern Mississippi, Claiborne found thousands of cattle grazing on three-foot high grass. The inhabitants of the woods were "for the most part pastoral, their herds furnishing their chief revenue. . . . Many of the men spend days in the woods herding cattle or deer stalking." Claiborne continued: "The most juicy and richly flavored grass-fed beef can be bought at three or four cents . . . and an owner of five hundred or one thousand head of cattle will thank you for penning, milking, and salting his cows."[12]

At Leaksville, just a few miles from the Alabama line in Greene County, Mississippi, Claiborne observed a piney woods roundup and skillfully recorded the event for posterity:

> Many of the people here are herdsmen, owning large droves of cattle, surplus increase of which are annually driven to Mobile. These cattle are permitted to run in the range or forest, subsisting in summer on the luxuriant grass with which the teeming earth is clothed, and in winter on green rushes or reeds, a tender species of cane that grow in the brakes or thickets in every swamp, hollow and ravine. The herdsmen have pens or stampedes at different points in the forest, where at suitable times they salt the cows, and once or twice a year they are all collected and marked and branded.
>
> This is a stirring period and quite an incident in the peaceful and somewhat monotonous life of the woodsman. Half a dozen of them assemble, mounted on low built, shaggy, but muscular and hardy horses of that region, and armed with raw hide whips of prodigious size, and sometimes with a catching rope or lasso, plaited of horsehair. They scour the woods in gallant style, followed by a dozen fierce looking dogs; they dash through swamps and morass, deep ravines and swim rivers, sometimes driving a herd of a thousand heads to the pen, or singling out and separating with surprising dexterity a solitary steer which has become incorporated with another herd. In this way, cheering each other with loud shouts and making the woods ring with the crack of their long whips and the trampling of the flying cattle, they gallop thirty or forty miles a day and rendezvous at night at the stamping ground.[13]

The focal point and market for much of southern Alabama's cattle industry was Mobile. The old city had established itself as a "cow town" in the eighteenth century and continued to offer a buying market for cattle raisers in the antebellum era. Cattle raisers drove cattle from the interior piney woods to Mobile. In the colonial and early antebellum periods, many cattle were shipped from there to the West Indies or other markets. Butchers slaughtered the rest to supply beef, leather, and other products to Mobilians and cotton plantations upriver.[14]

Cattlemen selling in antebellum Mobile did not always receive favorable prices for their stock. The story of a determined piney woods herder illustrates the point. In the fall of 1838, a cattleman from the interior drove his herd to Mobile, where butcher-merchant monopolies charged locals exorbitant rates for beef despite an abundance of cattle in the countryside. Refusing to sell his stock at seven cents per pound—butchers were selling beef for a quarter a pound—he set up his own shop outside city limits and sold his beef for one bit per pound. Within a week the concerned butchers bought him out at that price.[15]

Though cattle herding played a more prominent role in the culture and economy in the piney woods than in any other region of Alabama, farmers and planters of all areas depended on and raised cattle. In northern Alabama cattle raisers similar to those in the piney woods likely roamed the Tennessee Valley before being pushed out by cotton planters or settling to plant themselves. A Morgan County historian observes that in the early years of white settlement herders from the hills to the south drove their cattle to winter in the cornfields and cane brakes of the Tennessee Valley. In the 1820s, Adam Hodgson witnessed farmers and herders in the Tennessee Valley west of Athens ranging "large quantities of sheep, pigs, and cattle, in the woods, with no other trouble than putting a bell around their neck, and occasionally visiting them." Cotton quickly became the focus of Tennessee Valley agriculturists, but scattered herders and drovers survived. In 1839, one such drover wrote to C. C. Clay of Huntsville of his intentions to "drive 50 to 100 beeves" to a butcher in that town.[16]

Subsistence farmers and yeomen dominated the mountains and hills. The farmers of these areas often maintained a diverse array of animals and crops. In addition to planting corn, cane, vegetables, and sometimes cotton, these farmers raised cattle, sheep, hogs, and poultry. Hogs were the chief products of most livestock raisers in the hills and mountains. The mast available in oak-hickory forests provided winter forage for hogs ranging free on hillsides and in remote coves and hollows. Nevertheless, cattle remained important to hill farmers in Alabama.

The agricultural censuses reveal substantial livestock raisers in the hill country of northeastern Alabama. In 1850, William Bynum of DeKalb

County owned 154 head of cattle, 75 hogs, and 12 sheep on a 365-acre farm. In Randolph County, Stephen W. Herrin ranged over 500 cattle, 100 sheep, and 1,500 hogs and grew no cotton on his 950-acre farm. DeKalb County herder William Sloan offered a prime example of an antebellum livestock raiser. Though he owned only forty acres and grew no cotton in 1860, Sloan owned 246 head of cattle, 30 sheep, and 24 hogs.[17]

The open range allowed farmers like William Sloan to turn their stock out to graze in the countryside. In that way the open range served as a democratizing element in antebellum Alabama. For many planters, however, the open range simply provided an obstacle to improving their herds. Concentration on cotton planting, the open range, and a lack of incentive prevented breed improvement in Alabama in the first half of the nineteenth century, though it was well under way in the North. After scattered attempts toward breed improvement before 1850, the decade before the Civil War witnessed a blossoming movement in Alabama to import purebred stock for breed improvement. Naturally, this took place among Alabama's wealthiest agriculturists, the planters of the Black Belt and Tennessee Valley.

Most Alabama planters had come to their new homes with their livestock as an essential part of their households. Cattle, especially, performed a number of diverse and necessary tasks for the frontier farmer. Before James A. Tait left Georgia for his Alabama River plantation in Wilcox County in 1824, he and his father had owned 150 head of cattle. The winter of 1824 and 1825 and the family's need for food likely depleted his herd, which numbered only thirty-eight in the spring of 1825. Tait used beef for his family and his numerous slaves; therefore, by 1843, he had rebuilt his herd to 160 head.[18]

The number of livestock differed from plantation to plantation and often depended on the preferences of the owner. Black Belt planter Hugh Davis preferred to purchase his beef. He owned only five cows, four calves, and two yearlings in 1851. According to Sam Bowers Hilliard most southerners preferred pork, which was easier to preserve. Davis's hogs and sheep far outnumbered his cattle holdings. John Park, a Pike County planter, frequently mentioned killing hogs, but never cattle. Even Barbour County planter John Horry Dent, who frequently had steers or heifers slaughtered in the spring, focused on efforts of planters to raise more hogs in an 1848 letter to the patent office.[19]

Other planters like Tait owned considerable herds of cattle. In 1860, Francis Merriweather Gilmer, who owned five plantations in Lowndes and Montgomery counties, owned some 200 head of cattle, 213 sheep, and 884 hogs. Marengo County planter Gaius Whitfield owned more than 7,000 acres, 213 head of cattle, 145 sheep, and 625 hogs. Several Tennessee Valley planters also had substantial livestock holdings. On his 6,620-acre Lauderdale County plantation, John Peters possessed 116 cattle, 150 sheep, and 960 hogs in 1860. Another Lauderdale County planter, G. G. Armistead, maintained 85 cattle, 200 sheep, and 250 hogs on his 3,000-acre plantation.[20]

On Black Belt and Tennessee Valley plantations selected slaves often served as cowhands. John Horry Dent referred to slaves minding his stock. In the spring of 1856, Dent supervised the building of a new cowpen by his slaves. Slaves participated in roundups on the plantation of James A. Tait, and in 1825 Tait fired an overseer for "involving my negroes with his own, in the killing of and in work in beefs." Robert W. Fogel and Stanley L. Engerman assert that the raising of livestock, including the growing of feed, occupied one-fourth of a slave's labor time, almost as much as working cotton.[21]

Slaves often managed the moving of cowpens on plantations. Cowpens had been used since the early days of cattle raising in colonial Carolina, and herders and planters found several uses for them as well. On the frontier and in the backcountry herders often used movable two- or three-acre cowpens to collect and keep selected milk cows at night, ensuring the family of a steady supply of dairy products. Agriculturists practiced the same system, though their primary goal was the fertilization of sections of fallow land. Slaves or hired hands dismantled and relocated pens every two or three weeks to replenish cotton lands. Cowpens took on an added role in the late antebellum period when large planters began to purchase purebred British stock. Some planters, such as John Horry Dent, kept their prized cattle in pens to prevent breeding with poor native stock. Dent advised his overseer, "Keep your Bull at home in a pasture; Gentle, well fed, salted, and frequently handled."[22]

In a 1944 article, James C. Bonner notes that some Alabama planters began improving their livestock in the late antebellum period. Bonner also states that many planters purchased additional lands on which to graze livestock. Nevertheless, breed improvement became popular among Alabama planters only in the 1850s. Consequently, according to Lewis C. Gray, most Black Belt planters bought their beef throughout most of the pre–Civil War era, and Sam Bowers Hilliard termed both the Tennessee Valley and the Black Belt "beef deficient" areas on the eve of the war.[23]

In the 1850s, while the herders and drovers retreated farther into the inner frontiers, some of central Alabama's prominent planters began importing purebred stock and attempting to improve the quality of their herds. The importation of improved, thoroughbred cattle from Europe, primarily Great Britain, dates to the Revolutionary era. Registered, purebred British stock remained scarce in the South throughout the antebellum period, however. By 1859 only 106 Shorthorn (Durham) bulls had been registered in the South.[24] While Shorthorn blood would definitely have improved Alabama's native stock—native stock consisted of a blend of Spanish, British, and probably French breeds—this breed's meager milk production dismayed most concerned planters. Unlike piney woods herdsmen, cotton planters desired improved breeds of milk cows, and since plantation owners were often the only agriculturists possessing the means or the desire for herd improvement, Alabama's improved stock types primarily consisted of those rendering greater quantities of milk.

In the 1850s a number of Alabama's elite, Black Belt planters became interested in improving their milking stock. Dissatisfied with Herefords and Shorthorns, which "are not the things for a cattle country," these agriculturists began to seek other breeds. Southern agricultural publications of the time denounced piney woods cattle that "do not give a quart of milk a day" and lamented the common occurrence of "planters with large herds hav[ing] to buy butter and go without milk." Despite one historian's claim that no enthusiasm for the Devon breed developed, by December 1856 Alabama's Noah Cloud reported that "this breed [Devon] is rapidly gaining favor in our state." In late 1853 Greensboro planter Isaac Croom imported a young Devon bull from Atlanta, and Bolling Hall led a group of Autauga County planters in the importation of a number of Devon cattle. In the winter of 1855 three Montgomery planters, Dr. William O. Baldwin, Samuel Swan, and Thomas B. Brown, purchased Devons from Richard Peters of Atlanta, one of the nation's leading Devon breeders. Alabama planters favored Devon cows for their production of fifteen to twenty quarts of milk per day; this was twice the average daily production of the Alderney, later called Jersey, breed.[25]

Though Alabama's planters and agricultural press gave the Devon top billing in the 1850s, many continued to import and raise Shorthorns and other breeds. In January 1854 one Black Belt planter reported the acquisition of two Shorthorn bulls and two heifers through an agent in Poughkeepsie, New York. The same planter claimed to have imported a bull directly from England as early as 1836. He wrote, "The improvement in cattle, and the cattle of my neighbors, from being bred to him for so long a time, is most manifest!" In the same year, the *American Cotton Planter* of Montgomery called for names of those wishing to pay $350 a head for the importation of a Swiss milk cow.[26] Cattle categories at the first state fair of Alabama in 1855 included Devons, Shorthorns, Ayershires, mixed breeds, and natives.

Breed improvement did not come without costs, however. At the 1856 state fair sponsored by the Alabama State Agricultural Society eight-month-old Devon calves sold for $200. By 1858 Devon bull calves were selling for as much as $250, while heifers often brought over $300.[27] Such prices confined breed improvement to the wealthy planters of the Black Belt and Tennessee Valley and the wealthy stock owners around Mobile.

The controversy over the open range also reflected class and wealth implications. The question of closed versus open range, which ultimately became the single most important local political issue of the late nineteenth century in the South, had its beginnings in elite agricultural thought before the Civil War. Echoing the sentiments of the South's foremost antebellum agriculturist, Edmund Ruffin of Virginia, leading Alabama planters and plantation advocates clamored against their state's traditional open-range system. Among these were publisher Noah Cloud and Marengo County planter Isaac Croom, whose essay advocating the closing of the range won first prize at the West Alabama Fair at Demopolis in 1859.[28]

Though most prominent planters maintained large herds of cattle, many resented the use of their grazing and forest lands by the herds of landless whites and small farmers. These stock law advocates disliked the necessity of fencing in their crops, which often grew on hundreds of acres, to keep out the poor stock of their less fortunate and less industrious neighbors. Such planters owned adequate grazing lands for their own stock, and many realized the benefit of "soiling," or periodically moving cowpens from place to place to enrich the soil with manure. Furthermore, those planters attempting to improve bloodlines in their own herds frequently found the open range a formidable deterrent. Many found themselves compelled to fence in their Devons or Shorthorns to prevent breeding with inferior native cattle.

The preceding pages offer a glimpse into the diversity of cattle raising in the state as well as a look at some trends in the antebellum era. Backwoods herders persisted throughout the period, and a more modern form of cattle raising began to take shape among the planters of the Black Belt. Nevertheless, the popular, historical view of the antebellum era as a time of rapid cotton expansion still merits respect. Though the numbers of cattle in Alabama increased in the twenty years before the Civil War, they pale in comparison with cotton figures. Between 1840 and 1860 the number of Alabama cattle recorded by census takers increased more than 15 percent, from 668,018 to 773,396. Despite the growth in raw numbers, the ratio of cattle to people declined significantly, from 113 head of cattle for every 100 persons to 81 head for every 100 persons. During the same period cotton production increased more than threefold, from 292,847 to 989,955 bales per year, or from roughly one-half bale per person in 1840 to more than one bale per person in 1860.[29]

Frank L. Owsley observes the decline in antebellum herding and finds the decline in a combination of herders converting to planting and additional herders being pushed into the inner frontiers. He writes: "With such great landed resources so cheap and available and such a relatively small population, it was inevitable that the majority of the agricultural population, and even those dependent upon a grazing economy, should become freeholders in the newer portions of the Old South." Those not willing to settle as yeomen or planters retreated into the piney woods or mountains where they "found sanctuary from the pursuing agricultural settlers." Owsley probably overstates the frequency of both cases. Census figures reveal a significant number of herders in the late antebellum period who produced no marketable crops. Furthermore, as Terry G. Jordan and John D. W. Guice have demonstrated, many of these cattlemen had inhabited the piney woods by choice, not out of necessity. Despite these slight exaggerations, Owsley's basic thesis regarding a declining and retreating cattle culture and industry appears quite applicable to antebellum Alabama.[30]

Because of the expansion of cotton, the retreating herdsman culture, and the effects of attempts toward improved breeding and planter diversification, one finds a number of changes evident in Alabama's cattle industry by 1860. Between 1840 and 1860, according to census returns, numbers of cattle decreased in the traditional cattle-raising piney woods of southern Alabama, especially in areas of expanding cotton cultivation. The cattle population of Covington County declined from 14,173 to 10,007 while that of Conecuh County fell from 22,269 to 15,748. In the same twenty-year period cotton production in Covington County grew from 759 bales to more than 2,000; Conecuh County experienced an even greater expansion of cotton production, from 1,750 bales in 1840 to 6,850 in 1860.[31]

Even more significant was the dramatic decline in cattle numbers per farm in southern Alabama. Between 1850—when the census first listed the number of farms in the state—and 1860 the average cattle herd on Alabama's farms declined from seventeen to fourteen head. The piney woods counties experienced the most dramatic drops in the ten-year period. While several Black Belt and Tennessee Valley counties experienced increases in the 1850s, Washington County's average holding plummeted from ninety-three per farm to nineteen. Average herds in Covington County dropped from seventy-seven in 1850 to nineteen a decade later. Baldwin County, which maintained the highest per farm cattle average in both years, witnessed a decline from 115 to 81 head. These significant decreases denote both the influx of yeoman agriculturists and a gradual conversion to planting by some former herders.[32]

Cotton could be found in some backwoods areas by the late antebellum period. As Frank L. Owsley suggests, some herders eventually settled into a planter or yeoman life-style. And, according to John Solomon Otto, "as newcomers swelled the backcountry population, more lands passed into private hands, and cultivated lands encroached on the unfenced range." For instance, in Randolph County the most prominent herder of 1850 had become a small planter by 1860. Stephen W. Herrin, who raised some 515 head of cattle, 1,500 hogs, and no cotton in 1850, altered his agricultural practices over the following decade. In 1860, Herrin produced thirty bales of cotton and owned only 220 head of cattle and 500 hogs. In Baldwin County,

Joseph Booth became a substantial planter between 1850 and 1860. Booth's cotton production soared from no bales in 1850 to 132 in 1860, while his cattle herd decreased from 260 to 159 head. A less dramatic switch occurred in Washington County, where Thomas G. Smith increased his cotton production from no bales in 1850 to thirty in 1860. Though Smith's hog count declined from 300 to 60, his increase in cattle numbers from 304 to 400 reflected the area's continuing reliance on cattle raising.[33]

Though some piney woods cattlemen entered the cotton market and others moved, several remained with smaller herds. According to manuscript census records the ten leading cattlemen of Covington County in 1850 owned an average of more than 250 head, and eight owned more than 200. Ten years later the county's top ten cattle owners possessed an average of fewer than 175 head, and only three owned more than 200 head. In Washington County, however, and most specifically in the St. Stephens area, where cotton remained scarce in 1860, cattle herds among top cattlemen increased slightly between 1850 and 1860. In 1850 ten cattlemen owned an average of 438 head, and at least a dozen in the county maintained herds in excess of 200 head. In 1860 sixteen Washington County cattlemen owned over 200 head while the average herd among the ten largest herdsmen had increased to 465.[34]

In addition to the expansion of the cotton kingdom, several other factors threatened the survival of antebellum cattle herding. Antebellum southerners generally favored pork, which was believed to be nutritionally superior to beef. Sam Bowers Hilliard writes, "It is unlikely that beef was eaten regularly, especially among the poor whites." Because of the difficulty of preservation, beef was commonly eaten fresh. When "a beef" was slaughtered the meat was often divided among neighbors or, on a plantation, among the slaves. Though the hill country of the South had a stronger dependence on cattle for subsistence and plantations often maintained herds of up to 200 head, according to Hilliard, the average southerner ate only twenty-five to thirty pounds of beef a year before the Civil War. By contrast, slaves consumed about 150 pounds of pork annually.[35]

The shifting southern frontier also affected Alabama's cattle fortunes. In the decade before the Civil War Texas became the undisputed king of the southern cattle range. By 1860 Texas contained an average of over five and one-half head of cattle for every one person, and this youngest southern state accounted for more cattle than Georgia, Alabama, Mississippi, and Louisiana combined. The increasing numbers of Texas cattle reaching St. Louis, Chicago, and New Orleans, combined with a growing demand for "improved" fatty beef in the Northeast, undoubtedly injured Alabama cattle herders, whose scrawny native stock rarely reached a mature weight of more than 800 pounds and usually yielded a carcass weight of about 300 pounds.[36]

The expansion of cotton planting in southern Alabama limited cattle ranges, and a growing urban population in Mobile consumed increasing amounts of local beef. In his study of food supply in the antebellum South, geographer Sam Bowers Hilliard notes a decline in Mobile beef exports after 1850 and a concurrent rise in beef imports. Though cattlemen continued to sell their stock in Mobile, most animals were slaughtered and consumed in town or shipped upriver to cotton plantations. An 1855 report of Mobile exports underscores the growing chasm between cotton and cattle in Alabama. In that year merchants shipped by river or by sea some 22,000 hides worth $55,000 and 8,800 pounds of tallow at ten cents per pound. Furthermore, no mention was made of the exportation of live cattle or beef. Conversely, cotton shipments amounted to over one-quarter of a billion pounds worth $22.8 million. By 1860, it was not unusual for ships to deposit hundreds of cattle per week in Mobile.[37]

The decades preceding the Civil War in the Black Belt, on the other hand, marked an increase in the cattle population among planters. The Marengo County cattle population increased by almost 50 percent, from 15,126 to 22,514 between 1840 and 1860; similarly, Dallas County counted 10,379 head of cattle in 1840 and 16,394 twenty years later. The Black Belt's largest cattle owners were not Owsley's yeomen or McWhiney's and McDonald's Celtic herders. In 1850 only one of Lowndes County's top ten cattlemen raised fewer than forty bales of cotton and had fewer than 350 acres under cultivation. By 1860 that county's typical top cattle owner maintained over 100 head of cattle in addition to cultivating 1,300 acres of land and producing more than 275 bales of cotton. The eve of the Civil War found the antebellum livestock herder absent in the Black Belt.[38]

Cattle numbers in northern Alabama generally remained consistent in the two decades before the war. Cattle herds in the Tennessee Valley declined somewhat, but increases in the mountainous areas offset the decline.

The valley's planters had long taken advantage of stock from Tennessee and Kentucky and in some instances cattle and hogs from the mountain counties of Alabama. According to one Tennessee Valley citizen, though, in the five years preceding the war area planters had attempted to increase livestock holdings in order to raise their own meat supplies, just as Black Belt planters had done for a decade or more.[39] Nevertheless, by 1860 only one Lauderdale County planter owned more than 100 head of cattle and, as in the Black Belt, all the largest cattle owners also produced substantial amounts of cotton. In the mountain counties, where the raising of cotton had made little headway by 1860, yeomen and herders continued to raise animals for subsistence and occasionally for market. In DeKalb County in 1860 only five farmers reported at least 100 head of cattle, and none of these raised as much as two bales of cotton.[40]

The transformation and decline of Alabama's cattle industry was under way by 1860. Herdsmen, planters, and yeomen faced four years of war that would severely deplete the livestock population. Between invading Union armies, local Confederate units, and domestic consumption, Alabama's agricultural families would lose a significant percentage of their prewar stocks of cattle, swine, and sheep. The ravages of war would leave many destitute and accelerate the existing forces influencing the steady demise of Alabama's traditional, open-range cattle industry.

King Cotton, westward migration, and efforts for breed improvement all affected the cattle industry and cattle raisers in antebellum Alabama. The piney woods cattle business was stagnant if not in decline by 1860, and Black Belt planters appeared more interested in milk producing than beef raising. The Civil War would certainly devastate the cattle industry in certain parts of the state, but the groundwork for decline had been laid. The future of the cattle industry lay to the west on the plains of Texas. Alabama would be cotton country.

Although the Civil War does not deserve sole credit for destroying the southern herding culture—indeed, fragments of the herding culture would survive well beyond 1900—the conflict undoubtedly played an important role in its decline as well as in the destruction of much else in southern society. The war wrought terrific change in every facet of southern life. The abolition of slavery raised difficult economic and social questions that demanded solutions. Defeat left hundreds of thousands dead, and military occupation bred animosity among the surviving millions. Many, especially in the most war-ravaged areas of the old Confederacy, found themselves destitute and starving, stripped of worldly possessions as well as dignity. At the heart of all these problems lay the near destruction of agriculture, the South's economic and social lifeblood.

Observers during and after the war continually commented on the "very destitute condition" of occupied lands and their inhabitants. As early as the summer of 1862 a Confederate soldier near Chattanooga observed "ragged and Dirty and bar [sic] footed" women and children who "look like that they did not get enough to eat." A Coosa County man, after having made a journey into northern Alabama in the spring of 1864, wrote, "No man who has not gone through the interior and examined into the condition of things knows anything of the absolute want and destitution among that class of people . . . " J. T. Trowbridge, a postwar chronicler of southern desolation, found Selma a "scene of 'Yankee vandalism' and ruin."[41]

The Civil War took an immense toll on Alabama's production of livestock and crops. Utilizing the less than precise figures of census returns before and after the war one finds a decline of more than one-third in the number of horses and mules between 1860 and 1870. Likewise, the number of sheep declined by more than one-third, from 370,156 to 241,934. The most drastic reductions, however, were in the numbers of cattle and swine. The number of hogs reported by the census in 1870 marked a decrease of over one-half from the 1860 count. Alabama's cattle population plummeted from more than three-quarters of a million in 1860 to less than a half million in 1870.[42]

After the war many southerners recounted with dismay the destruction of their antebellum, "diversified" agricultural society, though the perception of this society and time was naturally clouded by the passing of years and the despair of defeat. Alabamian John T. Milner decried his state's desolation and agricultural transformation. Describing a rich and highly diversified antebellum agricultural economy, Milner compared the state's most productive Tennessee Valley and Black Belt counties with corresponding counties in the Midwest and far west. He found a degree of antebellum diversity of production in Alabama capable of competing with that in other premier American farming sections; naturally, the postwar picture of

Alabama's farms, with "no fences, no hogs, no cattle, no agriculture, no nothing," could not hope to measure up to its counterparts elsewhere.[43]

Recent scholarship on the transformation of southern agriculture has similarly highlighted the four years of war and the early years of Reconstruction to the virtual exclusion of the half century of change preceding 1861. Grady McWhiney finds a sharp change in the nature of Alabama's agrarian society between 1860 and 1870, and he notes that the Civil War and Reconstruction were particularly "devastating to Alabama's animal growers." According to McWhiney and Forrest McDonald, "The capital losses in the form of animals destroyed by both Confederate and Union troops during the war were as important to the drovers, relatively, as were the capital losses to the planters that resulted from emancipation."[44]

In Alabama, as we have seen, the fifty years preceding the war witnessed a rapidly expanding cotton culture pushing once prevalent herders and drovers into the "inner frontiers"—the southern piney woods and the mountainous and hilly sections of northern Alabama. By 1860 Alabama farmers and planters raised cotton in all fifty-two counties, and only two counties produced less than 1,000 bales. Though large planters, especially those in the Black Belt, continued to maintain substantial droves of swine, the traditional hog drovers had largely retreated to the mountains by 1860, where oak-hickory forests supplied plentiful mast for their animals. Likewise, the piney woods offered a last haven for cattlemen in 1860. Among cattle raisers we find an example of a declining culture, pressed by geographical and economic forces in the antebellum period and hastened toward extinction by the Civil War's destruction. McWhiney and McDonald justifiably stress the Civil War's impact on southern agriculture, but in their attempts to identify the cultural transfer of Celtic herding tradition they overlook important antebellum agricultural trends in Alabama and elsewhere that had begun to undermine the importance of cattle herding prior to 1860.[45]

Contemporaries and historians have stressed the South's abundance of livestock at the beginning of the Civil War. According to figures published by the U.S. Commissioner of Agriculture in 1863, the South "as a whole was overstocked with cattle" in 1860 with a surplus of over one and a half million head. The same study estimated Alabama's surplus cattle population at approximately 125,000. Historian Paul Gates notes the region's "great supply of hogs and cattle," though stock was inferior except in the upper South. Historian John Solomon Otto, in his recent study of southern agriculture in the Civil War era, estimates that the Confederacy contained twice as many cattle as it required for subsistence in 1861. Nevertheless, the vast majority of this cattle surplus ranged in Texas, far away from Confederate army demand, and east of the Mississippi only Florida possessed a substantial surplus cattle population.[46]

Several factors contributed to the depletion of Alabama's cattle population during the course of war and reconstruction. Though depredations and impressment by invading armies have received credit for much of the loss, and rightly so, even more cattle were consumed by Confederate forces. Confederate officials acquired stock and cured meat through a variety of means, including purchases by civilians or army officers, impressment, and the tax-in-kind. Unauthorized impressment and theft by roving bands of Confederate and Unionist marauders, as well as domestic consumption and disease, also reduced herds throughout the state.

The four years of war saw federal and confederate operations eventually reaching most sections of the state. Northern Alabama suffered most severely with three years of intermittent federal occupation and constant pressures from southern troops and raiding cavalry units. Central and southern Alabama were spared the harsh effects of Union operations until late in the war, though the Confederate garrison at Mobile and traveling commissary agents and impressment officers constantly required supplies of all kinds from planters, small farmers, and herdsmen alike.

Beginning in early 1862 federal raids began to harass residents of northern Alabama's Tennessee Valley. These first invaders found a prosperous farming country with a significant number of Unionists. Though the valley witnessed no major battle, Union Major General Ormsby Mitchel maintained his headquarters at Huntsville throughout the summer of 1862 before abandoning northern Alabama in the wake of Braxton Bragg's invasion of Kentucky. Union forces returned to the Tennessee Valley in February 1863 where they maintained tenuous control, except for a brief time in 1864 before the Battle of Franklin, until the end of hostilities.[47]

The citizens and agriculture of northern Alabama suffered terrible losses during the Civil War, and the Tennessee Valley bore the brunt of this suffering. Walter L. Fleming notes, "In the counties of Lauderdale, Franklin, Morgan, Lawrence, Limestone, Madison, and Jackson nearly all property was destroyed... All farm animals near the track of the armies had been carried away or killed by the soldiers." A May 1862 U.S. Fifth Division circular concerning subsistence in the Tennessee Valley read: "All livestock in our lines must be driven in and used, and all grass, wheat, and everything fit for forage gathered." Another Union officer of the 1862 campaign testified that "everything that could be taken without absolutely starving the women and children was taken," and still all the foragers could obtain were "a few almost worthless cattle and sheep."[48]

Though the Union army initially attempted to supply a large portion of its troops in northern Alabama with beef cattle driven or transported by train from Nashville and Louisville, bands of Confederate guerillas often captured cattle and always made it difficult for beef contractors to deliver their stock safely. Consequently, Union command grew less reluctant to live off the supplies of occupied lands. General Don Carlos Buell ordered his army's foraging parties to bring in "all the cattle for many miles around... milch-cows and yearlings and everything [they] could scrape up." Lieutenant Francis Darr and four companies of cavalry scoured the banks of the Elk River from its mouth to Fayetteville, Tennessee, and returned "with 200 head of cattle, which was every hoof which could be gathered up on both banks of the river." Another Union officer recalled that "we would send out into the country and get in beef cattle frequently and slaughter them... [I]n one instance we had a yoke of oxen sent to me, that were captured by one of our scouts, belonging to a Confederate officer. They were very poor and thin, but I was ordered to turn them over to the commissary to be killed."[49]

Union troops often found the Tennessee Valley's citizens and stock in poor condition. Major General Lovell H. Rousseau discovered great want of cattle, hogs, and other provisions in late 1862. Noting that "North Alabama is not a cattle country," Rousseau found "inferior cattle, small and thin." Though his command consumed hundreds of animals and "several hundred... were sent up to the troops at Stevenson on the cars," Rousseau believed they had left northern Alabama "without using all the cattle while there." Furthermore, the general felt that the region's cattle supply had been devastated by the initial occupation of Mitchel's troops, since Mitchel's command suffered communication problems with its supply base and thus had to rely heavily on local provisions. General W. S. Smith supported this claim by stressing the disrepair of the Louisville and Nashville Railroad during Mitchel's stay in northern Alabama. General Smith believed "the inhabitants of Huntsville were actually in want" by the time Mitchel left in August.[50]

One local Unionist offered an explanation for the valley's scarcity of stock by late 1862. Before the war planters maintained just enough livestock to live on, if even that much, because cotton planters had "come to the conclusion that the raising of cattle [would not] pay." After the fall of Fort Donelson retreating Confederates had levied supplies from valley citizens, and subsequent Union occupation had simply depleted the bulk of stock remaining.[51]

Union troops in the Tennessee Valley in 1862 also committed perhaps the harshest atrocities on people and their belongings reported in Alabama. Though General Orders Number 81 of the Third Division headquarters specifically prohibited all plundering, pillaging, and depredations upon property, murders, rapes, and thefts were committed by "lawless brigands and vagabonds connected with the army." In May 1862 a group of forty-five Athens citizens reported an aggregate loss of $54,689.80 due to Union depredations. During the occupation of Huntsville ex-governor Clay of Jackson County claimed that Union soldiers repeatedly robbed him of stock and other things. Such losses were only repaid if acquisitions were legally made by authorized officers, and in many cases unauthorized troops gathered more provisions than their authorized officers. According to Rousseau, "Many of the soldiers had been very much in the habit of taking everything they wanted and many things they did not want." Lawless Confederates also terrorized the Tennessee Valley. The *Montgomery Advertiser* in late 1862 reported the existence in northern Alabama of roving bands of cavalry claiming to be Confederate troops who seized livestock and other provisions.[52]

Union and Confederate armies continued to provision themselves off Tennessee Valley lands throughout the war. In May 1863 Union Brigadier General Grenville M. Dodge reported in a raid of northwestern Alabama that his army destroyed 1.5 million bushels of corn and 500,000 pounds of bacon while capturing "1,000 head of horses and mules, and an equal number of cattle, hogs, and sheep." Dodge also "took stock of all kinds that I could find, and rendered the valley so destitute that it cannot be occupied by the Confederates." One year later a Union boat sailed down the Tennessee River from Chattanooga into northeastern Alabama where the crew landed and killed all livestock near residents' houses.[53]

Suffering perhaps even greater hardships than the Tennessee Valley were the mountainous counties of northern Alabama. In addition to enduring Union occupation and playing host to roaming Confederate forces, the mountains became a favorite haven for lawless groups of mounted men, bushwackers, and murderers. The mountain country's populace of subsistence farmers and herders fared poorly amidst conditions of war and occupation. Whereas the loss of twenty or thirty head of cattle and a similar number of hogs caused a planter considerable inconvenience and financial hardship, the loss of just a few cows, hogs, or sheep might threaten a mountain family with starvation, especially with the husband-father gone to war.

In August 1863 a Confederate commissary in Chattanooga reported no fresh beef on hand and "gloomy" prospects for bacon and beef after September. He expected to receive from northern Alabama not "more than 600 or 700 head—less than three days' supply." Five months later General Joseph E. Johnston in Dalton, Georgia, complained that General Leonidas Polk's smaller force was enjoying the provisions of Mississippi, western Tennessee, and the productive part of Alabama while his army had only an "exhausted country" from which to supply itself, that is, mountainous Alabama and Georgia. In March 1864 Brigadier General James H. Clanton wrote from Gadsden: "The country east, west, and north of this point to the Tennessee River is about exhausted in the way of forage, and the impressments heretofore have reduced the supplies of many families to less than is absolutely necessary for their maintenance for the present year." Nevertheless, less than a month later generals Johnston and Polk sent parties "into northeastern Alabama to procure beef." Depredations by Confederate troops also increased as the war continued in northern Alabama, where Clanton noted, "Our own cavalry has been a great terror to our own people . . . Stealing, robbing, and murdering is quite common."[54]

In the spring of 1863 Confederate commissary agent W. W. Guy scoured mountainous northern Alabama for cattle. Writing from Oxford Guy informed Colonel Benjamin S. Ewell that he had a herd of 80 to 100 cattle and hoped to secure as many as 600 within the week. Guy found slim pickings in the region, however, and soon retreated to central Alabama to continue his search. In the meantime, federal and Confederate impressment, stealing, and the tax-in-kind took a severe toll on northeastern Alabama. In April 1864 Governor Thomas H. Watts declared DeKalb County "distressed and destitute." Three months later the tax collector for the northeastern district suggested that collection be suspended and remarked that Cherokee County had been reduced to a famine state.[55]

Unlike the Tennessee Valley and the mountains of northern Alabama, southern Alabama remained almost totally free of Union invasion for the duration of the war. Small raids into Geneva and Covington counties from Florida at different times during the war and an operation by federal forces to sever the Montgomery and Pensacola Railroad at Pollard in early 1865 marked the only Union activity in southern Alabama until the fall of Mobile after Appomattox.[56]

Troops and civilians involved in the protection of Mobile undoubtedly drew on the surplus of beef in the vicinity. In 1861 the five top cattle-producing counties on the state comptroller's tax list were all in southern Alabama. In May 1863 A. D. Banks at Mobile offered to send 2,000 to 3,000 cattle to the Army of Mississippi on short notice. Three months later Mobile's acting commissary of subsistence reported a thirty-four-day ration for 20,000 men in the commissary depot. This included 347,410 pounds of fresh beef and 9,745 pounds of salt beef, in addition to almost 100 tons of hay and fodder for cattle. Even as late as January 1864 a Confederate officer at Mobile counted 400 head of beef cattle for the coming siege and stated with confidence, "There is plenty of meat now being cured in Alabama, and some beef cattle are awaiting my call."[57]

Southern Alabama's isolation and the scarcity of either Union or Confederate troops in the region bred an atmosphere of rustling by both Unionists and parties sympathizing with the Confederates. In June 1864 Governor Watts received letters from southern Alabamians complaining of thievery. One Abbeville man wrote, "Within about ten days two raids were made by the Yankees and Tories, during which our people lost 39 young negroes, besides provisions, stock, salt [and other items]." In nearby Pike County a group of angry citizens complained of an organized band of deserters and tories who "took all the stock and provisions that they could find."⁵⁸

Union sympathizers often fell victim to Confederate or southern rustling. Confederate troops confiscated several head of cattle from Alfred Holley of Covington County, a Unionist who sold cattle to federals in Florida and frequently led Union foraging parties in southern Alabama. Another Covington County man complained to Governor Watts that "there has been a great number of the cattle drove off under the pretence of confiscation by persons who have no right or authority from the government."⁵⁹

An 1864 act of the Alabama legislature reflects the depletion of the piney woods cattle population and the growing tendency for herders to trade with the enemy. In December 1864 the legislature passed an act "For the Preservation of Cattle in the County of Washington." The legislation made it illegal to sell butchered beef to a steamboat officer, a railroad agent, or a ship's representative without purchasing a one-year, five-dollar license from the probate judge. The act, which carried a $1,000 fine for lawbreakers, further required that a person applying for a license had to obtain written recommendation from at least five "respectable stock raisers of the county."⁶⁰

Central Alabama remained free of Union invasion until the last year of the war. Two Union raids, both of which originated in the Tennessee Valley, plunged into the heart of the state, beginning with that of Lovell H. Rousseau in July 1864. Major General Rousseau's cavalry—ordered by Sherman to destroy a section of the valuable Montgomery and West Point Railroad—traveled from Decatur to Loachapoka in seven days, living off the land as they went. After destroying a section of the railroad they followed the line through Opelika and into Georgia, destroying and capturing stores on their way. As historian Robert H. McKenzie has recognized, the brevity and swiftness of Rousseau's raid left little time for wreckage.⁶¹

Major General James H. Wilson's raid delivered considerably more damage on previously untouched parts of Alabama. In late March 1865 Wilson's unit left the Tennessee Valley and headed for central Alabama. At Elyton Wilson dispatched John Croxton's brigade to seize Tuscaloosa. Upon crossing the Cahaba River between Montevallo and Selma, the Union general found a country that contrasted starkly with the absolute destitution of northern Alabama, a region of abundant forage untouched by Union forces. Wilson's cavalry found in Selma a rich supply center for foodstuffs: beef, corn, bacon, hay, and other provisions. After taking and killing livestock and torching Selma, Wilson's men continued through Montgomery and into Georgia, often ranging foraging parties as far as fifty miles from the cavalry's main line. One Selma man said of Wilson's cavalry after the war: "They just ruined me. They took from me six cows, four mules, fifteen hogs, fifteen hundred pounds of bacon, eight barrels of flour, and fifty-five sacks of corn."[62]

Central Alabama towns from Demopolis to Opelika became supply centers for Confederate forces in Atlanta, Mobile, and Mississippi. In May 1863 Confederate commissary officer A. D. Banks informed Joseph Johnston that two Texas cattlemen were to deliver 5,000 head to Demopolis in the next sixty days. The same report stated that the depot at Demopolis already had about 2,000 head of cattle herded in the area or on their way.[63]

Because of the area's secure position throughout most of the war, central Alabama, and particularly the Black Belt, was found to be an ideal area for impressment. Impressment became the most controversial of several Confederate methods of collecting provisions. For the first two years of the war Confederate Commissary General Lucius Northrop operated an unwieldy system involving purchasing and collecting districts under chief commissaries, who often served as buying agents as well. Furthermore, states instituted similar systems to feed home guards and Confederate forces within their lines. By 1863 a number of factors forced the Confederacy to strengthen its methods of collection. Inflation ran rampant, and buying agents, who also competed with civilian contractors for livestock and other materials, frequently could not satisfy the financial demands of their suppliers. The dwindling supply of livestock and forage east of the Mississippi called for urgent action to secure much of the remaining stock for the armies. Most important, Union movements on the Mississippi threatened to eliminate the valuable supply of Texas cattle that had proved so important to the western theater.[64]

On March 26, 1863, Jefferson Davis signed an act regularizing and legalizing "impressments of forage, articles of subsistence, or other property absolutely necessary" by authorized officers. Paying with money or promissory notes, impressment officials were to make an offer to the supplier; if the supplier refused the offer, the property was to be appraised by "two loyal and disinterested citizens of the city, county, or parish," one chosen by the owner and one by the officer. The act also forbade impressments of bare necessities for subsistence.[65]

Under Northrop's impressment system the Subsistence Bureau appointed a chief commissary for each state—Major J. J. Walker in Alabama. The chief commissary then divided the state into districts, which generally corresponded to congressional districts, under a chief purchasing commissary. District chiefs could then further subdivide their districts and place subcommissaries over the smaller units. The plan also provided for the establishment of commissary depots in strategic locations. Northrop hoped that "when this system is thoroughly organized and worked, there will be no portion of the Confederacy which is not thoroughly drained, and, therefore, wherever our armies move, all the supplies of our country will be tributary to their use."[66]

Controversy soon arose over section five of the act, which provided for the appointment of state commissioners responsible for the joint arrangement and bimonthly publication of price schedules. From the beginning commissioners routinely underestimated prices and often went several months at a time before updating their low figures. The July 1863 price schedule set the price of beef cattle at sixteen to twenty cents per pound, depending upon the quality. Two months later the schedule stipulated eighteen cents a pound for all beef cattle. By September 1864 the impressment price for fresh beef had been increased almost threefold to seventy cents per pound, but the South's terrible inflation left a worthless currency.[67]

Within a short time of the inception of the impressment system, complaints arose from southern agriculturists over low prices and unauthorized impressment. In late 1863 the *Southern Cultivator* advised "those who have stock hogs or cattle by no means, if they can avoid it, to sell to butchers or army contractors." In the long run, the writer felt, the army and community would be better served with a more stable supply of livestock. Alabama's livestock owners frequently expressed their displeasure in writing to Governor Watts, who communicated these sentiments to Jefferson Davis and Secretary of War James Seddon. In January 1864 Watts wrote to Seddon: "It is a better policy for the Government to pay double price than to make impressments." Watts complained to Davis that "in a great many counties and instances, these Companies have done little good, and frequently, great harm." Furthermore, he informed the president that "frequently, too, two sets of officers make impressments at different times of the same persons. Thus irritation is produced, and opposition to the cause is generated." Again to Seddon he wrote: "Impressments, even when made by Confederate officers, are odious to the sense and feelings of the people."[68]

Unauthorized impressment became a particularly demoralizing problem for southern officials. In December 1863 Governor Watts warned Seddon of imposters "creating great disturbance and excitement" and dampening "the patriotic ardor of the People." Watts later complained to the secretary of war of impressments "made by indiscreet, and, frequently, heartless men." In June 1864 the Confederate Congress attempted to remedy the ills created by impressment with enactment of a temporary system of claims. The act allowed the secretary of war to appoint district agents to oversee claims made by citizens for impressment payments and ordered "payment of such claims as appear to them to be equitable and just." This system, which expired on January 1, 1865, satisfied few claimants. Consequently, the state of Alabama took measures of its own, passing an act in October 1864 outlawing unauthorized impressment and providing a two- to five-year sentence for violators. Furthermore, the Alabama legislature passed an act that required a fair market price to be asked for impressed goods.[69]

Depletion of the cattle population near armies and depots was particularly severe. The *Southern Cultivator* observed "much wanton destruction of milch cows and stock hogs near our large armies." In February 1864 the Confederate Congress amended impressment to protect milk cows and breeding stock on farms and plantations. Nevertheless, impressment of livestock continued until March 1865, when an act was passed making it unlawful "to impress any sheep, milch cows, brood mares, stallions, jacks, bulls, breeding hogs, or other stock kept or necessary for raising sheep, hogs, horses, mules, or cattle." The act further demanded that any impressments should be made at "the usual market price of such property at the time and place of impressment."[70]

The Confederacy also instituted a tax-in-kind in 1863. Eventually becoming as hated as impressment, the tax-in-kind, or tithe law, claimed for the government one-tenth of an enumerated list of goods. A state quartermaster and post quartermaster in each congressional district collected the goods at a specified depot. Though tax-in-kind legislation made no specific demands for beef, the tithe further depleted Alabama's exhausted resources. By June 1863 the destitution of fourteen Alabama counties made collection of the tax-in-kind impracticable within these areas.[71]

By the closing months of the war livestock supplies in the South were severely depleted. In October 1864 the Confederate Subsistence Department estimated its stock of bacon, pork, and beef would last 300,000 troops only twenty-five days. Daily rations had long been reduced from their initial, generous amounts. In February 1865 one southern publication lamented, "Our stock of neat cattle is getting greatly reduced, and not more than half the quantity of beef can be furnished in 1865 that was killed in 1864." The end of hostilities brought no end to hunger. Many planters and farmers claimed that freed blacks and destitute whites roamed the woods killing cattle for meat.[72]

Though there are no accurate postwar statistics for Alabama, it is safe to say the war significantly reduced the state's cattle population. The U.S. Commissioner of Agriculture estimated the number of cattle in Alabama in 1866 at approximately 410,000 as compared with the prewar census estimate of about 780,000. Four years later the 1870 census counted 487,163 head of cattle in the state.[73] The 1870 census reported only eight Alabama counties with more than 10,000 head of cattle, all but one of which were in the Black Belt or southern Alabama. Cattle numbers were lowest in the Tennessee Valley and in the mountains, where some counties reported fewer than 5,000 head. Impressment, pillaging, and occupation had claimed a significant percentage of livestock in the Black Belt and Tennessee Valley, both of which Sam Bowers Hilliard termed beef-deficient areas in 1860, as well as in the subsistence-farming areas of northern Alabama. Cattlemen in certain areas of the piney woods region of southern Alabama also suffered substantial losses, though the area entered the war with a much larger surplus of cattle.[74]

Historian John Solomon Otto utilizes statistics to downplay the Civil War's destructive effects on southern cattle raising. Otto correctly observes a more drastic decline in the number of hogs in the old Confederacy. However, he ignores much of the nonfrontier South when he writes that after 1870 "the region produced an abundance of beef." He speculates that this cattle increase may have occurred largely in Texas and Florida, but he presents only speculation whereas statistics bear out a certainty when examined in detail. Only Texas and Florida increased their cattle production between 1860 and 1870, and much of the next decade's growth occurred in these two frontier states.[75]

Whereas Otto, by looking at the whole South, tends to underestimate the Civil War's impact on cattle raising throughout the interior South and Alabama, Grady McWhiney and Forrest McDonald overestimate the conflict's role in the decline of Alabama's traditional, open-range herding culture. Though the Civil War certainly dented the state's cattle population and left farmers, herders, and planters in portions of the state devoid of a large number of their prewar livestock, the event more accurately acted in conjunction with antebellum trends already threatening cattle raisers. Although the traditional, open-range system survived into the twentieth century in certain isolated reaches of Alabama, the antebellum trends of cotton encroachment and class- and culture-based range struggles continued to push the once common herding and ranging culture into the backwoods and mountains. The Civil War was not the only factor changing "the nature of [Alabama's] agrarian society," but rather an important element, and often catalyst, of decades of cultural and agricultural transformation.[76]

Chapter 3

Agricultural Progressivism and the South

The half century between the Civil War and World War I was a period of social, political, and agricultural flux and change, readjustment, and adaptation. The Civil War and Reconstruction forced modifications of labor relations, political structures, financial and credit systems, and agricultural practices. Scientific and technological advances engulfed more and more of the South's and Alabama's nether regions in the Cotton Belt and threatened the traditional open range. The heralds of agricultural progressivism—public institutions, private organizations, and energetic individuals—encouraged and cajoled farmers and stock raisers to adopt modern, efficient methods. Pests injured crops and animals, leading many farmers to turn for help to agricultural reformers such as veterinarians, extension agents, and breed association representatives. As a result many farmers, especially in the Black Belt, would take the first steps toward a modern version of cattle raising. At the same time others would find that the progeny of progressive agriculture encroached upon traditional customs, threatened livelihoods, and demanded change.

The half century in question was one of stagnation, if not decline, for Alabama's cattle raisers. Nevertheless, despite the apparent inactivity among Alabama farmers and cattlemen and the tremendous spread of cotton cultivation, the foundation for the modern cattle industry was laid during this era. Though cotton production spread into every county of Alabama and the state's production of the fiber more than doubled, cattle persisted as integral subsistence and cash providers on farms and plantations. Furthermore, in large part because of growing urban demands for dairy products, the number of cattle in the state increased by more than one-third between Reconstruction and the onset of World War I (Table 1, appendix). A number of events and developments in this period directly or tangentially related to cattle raisers would prove to have even far more profound and lasting effects than market expansion on the future of cattle raising and agriculture in Alabama.

Though the modern cattle industry blossomed throughout the state only after World War II, the key blocks of its foundation were laid before 1920. The products of agricultural progressivism and the New South movement provided private and public bases from which individuals of influence could promote diversification and modern livestock breeding practices. From the establishment of Alabama's land grant college at Auburn in 1872 to the government-sponsored campaign for the eradication of the fever tick in the first quarter of the twentieth century, public officials and private individuals attempted to adapt technological and scientific advances and progressive ideals to the increasingly monocultural southern countryside. Though widespread realization of the progressive agriculturist's goals awaited New Deal intervention and World War II–era technological and economic upheaval, by World War I the first steps had been taken toward achievement of a cattle industry based on the midwestern model. This new system of livestock farming stressed the raising of purebred British breeds and the utilization of enclosed and improved pastures, winter forage feeding, veterinary care, and, on occasion, animal shelters. Simultaneously, the death knell began to ring for the few remnants of Alabama's traditional herding culture: open-range drovers, landless stock raisers, and semisubsistence farmers of the backwoods.

Despite the destruction and disruption of the Civil War, open-range cattle raising persisted well into the postbellum era. The extent and dispersion of this traditional agricultural practice differed according to geographic regions but qualitatively paralleled the antebellum regional model presented earlier. Open-range livestock raising flourished and endured primarily in backwoods areas of Alabama, the southern piney woods and rugged uplands in the north. Swine production continued to dominate in the hills and mountains; in southern Alabama many drovers maintained herds of cattle numbering in the hundreds along with large numbers of sheep and swine.

The manuscript agricultural census of 1880 reveals the persistence of several piney woods cattle raisers. The substantial number of postbellum livestock raisers forces us to reevaluate the emphasis of historians such as Grady McWhiney and Forrest McDonald on the central importance of the Civil War in the disappearance of southern open-range herding. Though the war obviously injured cattle raisers in the piney woods, the cattle industry was already in decline and retreat well before 1860. The postwar piney woods contained fewer cattle raisers and smaller herds; nevertheless, many of these families carried the herding tradition into the twentieth century.[1]

The expansion of cotton cultivation before and after the Civil War pushed herders into the more inaccessible and infertile reaches of the backcountry. In 1880, Baldwin County counted twenty-eight cattlemen and women with herds in excess of 100 head, and only one of those raised cotton. Likewise, in Washington County only two of the thirteen cattle raisers with more than 100 head raised cotton, and they combined to plant only thirteen acres. In DeKalb County none of the owners of more than fifty head planted cotton. Most herders raised varying amounts of other crops, including wheat, corn, oats, and cane. Furthermore, these agriculturists enhanced their subsistence efforts by utilizing selected cows for dairy purposes.[2]

The 1880 manuscript census schedules offer the historian a glance into the various usages of cattle in the open-range areas of the state. The 1880 census recorded the numbers of cattle bought, sold living, and slaughtered during the preceding year. Consequently, we can distinguish the methods of marketing and utilization. Marketing for several large herders in southern Alabama simply entailed driving a selected portion of their herd to Mobile, a railroad depot, or some other destination after a fall roundup. In Baldwin County seven cattle raisers sold twenty or more head from their large herds in the year preceding the 1880 census. Richard Moore marketed sixty head out of a herd of more than 300. Covington County cattleman Dennis Hart, one of the state's largest raisers, sold forty-six head from a herd of more than 1,000. Others relied on a higher percentage of their herd's natural increase. Baldwin County herder James Carney sold over one-third of his original herd of 115 head, and Elizabeth McMillan marketed thirty head, almost half her herd, in 1879.[3]

The livestock drover remained a valuable component of the small farmer-herder community in the late nineteenth century. The drovers, generally small farmers and stock raisers themselves, conducted annual treks to market, buying livestock from farmers and herders along their routes. Though drovers could be found throughout the state in 1880, census statistics reveal their stronghold to have been the hills and mountains of northern Alabama. James Hickson, a small subsistence farmer with a 160-acre farm in DeKalb County, purchased 200 head of cattle and sold 150. At the time of the census he owned only fifty head. James Heck, living in the same county, produced two bales of cotton and planted thirty-five acres of small grains and other crops and marketed 140 of the 158 cattle he bought. He was left with only thirty head at census time. Six other small farmers bought and sold twenty-five or more head of cattle, including Lawrence Smith, who owned only six cows in 1880.[4]

Three Winston County drovers are perhaps even more representative of the postbellum highland drover. Henry Warren, who produced one bale of cotton on a 460-acre hill farm, bought and sold twenty-five head in 1879, though he owned only ten head himself. Of the ten, Warren used two as work oxen and three for dairy purposes. He also slaughtered one for beef. Joseph Brumfield bought and sold ninety-one head in 1879, though he was obviously no large cattle raiser. He raised five acres of cotton and several other subsistence crops on his 350-acre farm. An even smaller farmer, Alonzo Sexton, sold fifty of the seventy-five head he purchased in 1879, though he retained only twenty head at census time.[5]

The practice of livestock droving was not confined to the uplands. In the Tennessee Valley, Madison County drover A. F. Steele and Lauderdale County drover Charlie Walk provide examples. Owner of a 265-acre subsistence farm, Steele purchased 100 head of cattle and sold ninety-nine. At census time he owned only one dairy cow and six other cattle. Walk, unlike most Lauderdale County farmers, planted no cotton on his 187-acre farm. He bought and sold 100 head of cattle in 1879, but he owned only eight at census time.[6]

In other parts of Alabama it appears that the traditional small farmer-drover had been replaced by larger farmer-stockmen or even planters. Nowhere was this more evident than in the Black Belt, a dark-soiled swath of land stretching across central and western Alabama and comprising some 4,300 square miles in parts of thirteen counties. In Marengo County five persons bought and sold over eighty head of cattle in the year preceding the 1880 census; only one of these produced fewer than ten bales of cotton. Frank McNeill, who owned almost 2,500 acres, purchased 100 head of cattle and sold 200. He still owned over 200 head at census time. Gaius Whitfield, Sr., one of the Black Belt's largest and most influential planters, bought eighty-seven head and sold 110, though he owned over 300 head at census time. Another large planter, J. S. Ryall, purchased 1,000 head of cattle and marketed over 1,400 in 1879. It is likely that many large planters purchased cattle from smaller producers in their communities. Many smaller Black Belt farmers reported sales but no purchases of cattle in the 1880 census. Large planters possessed the financial and material means not only to make purchases but also to transport the livestock to local butcher shops or railroad market centers. An example of a more traditional drover in Marengo County might be W. B. Kimbrough, a cash renter who raised thirty acres of cotton. Though he owned only fifty head of cattle in 1880, Kimbrough had purchased and sold 200 head in the previous year.[7]

Though many farmers and herders undoubtedly received cash from sales of live cattle, the most common uses of cattle in the postbellum era involved a number of subsistence practices. Throughout the state planters, yeomen, and tenant farmers milked cows, plowed behind oxen, and slaughtered an occasional "beef." Milk and butter production varied throughout the state. Among small farmers in Alabama, those in the northern sections tended to produce more dairy products, reflecting a degree of diversification unparalleled in other parts of the state. The bulk of this production was for home use, since commercial dairy activity was confined to urban areas.

Despite the southerner's penchant for pork consumption, most farmers slaughtered at least one "beef" each year. The lack of refrigeration and preservation methods generally forced immediate consumption of beef. Consequently, families in a community often alternated beef slaughterings and distributed the fresh meat throughout the neighborhood. Such springtime practices continued well into the twentieth century in many locales. Though a typical small farm household slaughtered fewer than four or five cattle annually, many large planters yearly butchered a dozen or more. These planters often distributed the meat among family members and tenant families. In Baldwin County only Richard Moore of the largest cattle herders slaughtered more than six head in 1879. Moore was also the only planter of the group, planting 150 acres in cotton. He likely divided his twenty slaughtered cattle among his tenants and workers. In Marengo County planters Gaius Whitfield, Sr., B. W. Whitfield, and Charles A. Poillnitz slaughtered ten or more cattle. In the Tennessee Valley, Lauderdale County planter James Angel slaughtered fifteen head of cattle on his 3,800-acre plantation.[8]

Planters like the Whitfields and Poillnitz maintained herds rivaling most found in the piney woods in the late nineteenth and early twentieth centuries. This fact reflected the continuation of pre–Civil War trends in the Black Belt, as well as the general increase in cattle numbers all over the state. Though the relative value of cattle to Alabama agriculturists declined as a result of the unbridled spread of cotton planting, Alabama farmers and herders increased their holdings by 40 percent between Reconstruction and the beginning of World War I. Nevertheless, the spread of cotton continually chipped away at the domain of open-range cattle herding and brought about laws and developments that would pave the way for cattle raising based on a midwestern model.

Nineteenth-century scientific and technological developments facilitated the expansion of cotton planting after the Civil War. The primary catalysts were railroads and commercial fertilizers. Railroads penetrated the "inner frontiers" in the last quarter of the nineteenth century, providing the markets and transportation necessary for large-scale cotton production.

Furthermore, superphosphates and sodium nitrate offered false fertility to some of the state's poorer soils. As a result, between 1879 and 1909 cotton production in former backcountry areas more than doubled. The Wiregrass and mountain counties showed the most dramatic increases over the thirty-year span. In the Appalachian region Cullman County experienced a phenomenal increase from 1,469 acres of cotton in 1879 to 42,338 acres in 1909. Less extraordinary but equally telling were increases in Winston County from 2,048 to 15,097 acres and in DeKalb County from 7,469 to 36,036 acres. In the Wiregrass region Covington and Geneva county farmers expanded their cotton production more than 1,000 percent, from a combined 9,123 acres in 1879 to over 99,000 acres in 1909. Surpassing these two in percentage increase was Escambia County, which enlarged its cotton acreage from 278 to 14,509 acres over the same period.[9]

The encroachment of cotton farming sometimes resulted in stagnation or a decline in cattle numbers. Though the state experienced an increase of more than 180,000 cattle between 1879 and 1909, a few Wiregrass and Appalachian counties failed to share in the growth. Counties such as Clay, Randolph, Talladega, Coffee, and Dale witnessed minor decreases in cattle numbers and a significant expansion of cotton cultivation during the thirty-year period.[10]

The simple fact that farmers increasingly broke their fields for cotton planting does not adequately explain the demise of open-range herding. A more immediate threat to the traditional practice was a product of progressive agriculturist thought containing elements of socioeconomic and class antagonisms—the movement to close the open range. The controversy over the open range began well before the Civil War in the South Atlantic states. By the late antebellum period some influential Alabama planters began to clamor for its abolition. Two of the leading advocates of such action were Noah Cloud, publisher of Montgomery's *American Cotton Planter and Soil of the South,* and Marengo County planter Isaac Croom. Cloud regularly ran editorials disparaging the open range, and Croom's essay on the topic won first prize at the West Alabama Fair at Demopolis in 1859.[11]

In Alabama the open range remained intact until after the Civil War. Wartime destruction of fences and stealing by desperate freedmen and poor whites, combined with prewar concerns for livestock improvement, strengthened demands for stock laws, or laws creating closed-range districts. In 1866, the Alabama legislature took the first step in an eighty-five-year process that would eventually end the open range throughout the state. On February 20, the legislature created the Canebrake Agricultural District, prohibiting the running at large of stock in a heavily cultivated plantation district in four Black Belt counties: Dallas, Greene, Marengo, and Perry. Over the next four decades this act was amended to include other areas, and nine other such districts were created by the assembly, including the Warrior Agricultural District (1879) and the Sylvan Agricultural District (1891) in Tuscaloosa County.[12]

Two additional and more common methods of restricting the open range arose during the late nineteenth century. Both methods involved legislative delegation of power to county authorities. In one case the legislature invested power in a local body. For instance, an 1881 act authorized the board of county revenues or court commissioners in twelve counties, upon the petitioning of ten landowners, to close the range in a particular section or beat. Under the second method the legislature authorized a county authority to call an election on the question after the petition of a certain number of property owners.[13]

The three legislative methods cited above aided the closing of the range over large sections of Alabama. Nevertheless, by the twentieth century the open range persisted in many of the less-cultivated areas of the state. In September 1903, the legislature voted to turn over full authority and control of the stock law question to local boards of revenue or courts of county commissioners. The act authorized these local bodies to call for and conduct stock law elections once a majority of freeholders in one beat petitioned for an election. This act set the stage for the heated, local stock law battles in the rural, backcountry beats of Alabama in the early twentieth century.[14]

In his work on the social and political transformation of the Georgia upcountry, Steven Hahn finds the stock law elections of the 1880s to be among the "roots of southern populism." The Georgia legislature in 1872 had passed a law similar to the 1903 Alabama statute, thereby precipitating locally centered feuds some thirty years before the local option arguments occurred in Alabama. Though some Alabamians in the 1880s struggled with issues similar to those in upland Georgia, the fiercest stock law battles awaited the local option law. As Hahn observes, many small and landless farmers saw stock laws as attempts by the agricultural elite to cast petty producers into economic dependence. Large farmers and planters, backed by railroad companies, attempted to justify stock laws with arguments for soil conservation and agricultural improvement. Although stock law proponents more likely acted out of a combination of agricultural progressivism and economic selfishness than from motives of vindictiveness and economic obstruction, the results were, nevertheless, class and cultural antagonisms. These antagonisms were further flamed by growing dependency on cotton under an oppressive, postbellum credit system.[15]

The open-range controversy in Alabama involved the same fundamental socioeconomic dimensions reflected in Hahn's Georgia upcountry, pitting planters and successful farmers against yeomen and landless agriculturists. Nevertheless, the absence of a local-option law in Alabama until 1903 prevented fence law struggles from playing an active part in the growth of agrarian protest groups. The range question appears to have exercised little influence on the success of groups such as the Grange, Agricultural Wheel, and Farmers' Alliance in the late nineteenth century, though scattered local battles may have influenced the strength of these organizations in certain areas.[16]

Before 1903, Alabama's stock law battles generally occurred in the more densely settled and heavily cultivated areas. These regions contained more substantial planters and large farmers whose economic interests lay in cotton production. In the fall of 1884, the *Opelika Times,* like many other small-town newspapers of the era, served as a forum for the discussion of stock laws. A proponent of the stock law, fearing that the legion of poor and landless farmers would vote to keep the open range, argued that large landholders' votes should weigh more heavily in the beat three election. Three weeks later a Gold Hill farmer vehemently disagreed with the sentiment, claiming that the "poor man . . . has the same interest at stake." One farmer from beat four observed that the stock law "eminently [*sic*] benefits one class of citizens to the detriment of another." The range controversy even elicited words of poetry from one Lee County stock law proponent. Concentrating on the ill effects of the open range on livestock quality, the anonymous bard wrote: "The shotted cow and crippled hog could many a tale unfold / Of treatment ill they have received owing to fences old / Then root them out—rebuild anew and give our stock a home / Where they 'mid sparkling clear and pastures green may roam."[17]

Laws preventing the free roaming of livestock inside town limits also proved controversial. One contributor to the *Opelika Times* noted that the stock law in that town met with much disfavor. An observer of the stock law in Courtland (Lawrence County) related the paradoxical problem at the heart of the issue. He wrote that to many minds the stock law is "a hardship upon the poor who have no place to keep their stock, and they will be forced to sell them." Like thousands of proponents, however, he justified the law with the argument that "thousands of acres can be cultivated with little or no fencing."[18]

After the 1903 law established local elections, stock law struggles erupted in many areas where the open range predominated. Walker County provides a useful example of a county beset by tumult. In the winter months of 1912 and 1913 stock law battles raged in this Appalachian county in northwestern Alabama. The Jasper *Mountain Eagle* regularly posted notices of stock law elections for various county beats, and both sides used the newspaper as a forum, though the editor's penchant for derisive remarks toward the law's opponents revealed his pro-law stance. In January 1912 alone, the *Mountain Eagle* advertised eight different upcoming elections.[19]

The Walker County opponents of the stock law certainly viewed the struggle as a class issue. In February 1912, Whit Alexander informed readers that he refused to vote for a law that would injure the poor. One year later Elbert Tubbs, a farmer of beat nine near Oakman, declared the law a "bad thing for the common farmers." Tubbs further railed against landowning coal and lumber companies, which generally favored stock laws, and claimed that the poor would "have no place to keep their stock." Evoking the religious fervor of the time, Tubbs finally warned that "a man that will vote for a law preventing poor folks and widow women's stock from trespassing on their land will never get to Heaven." Conversely, pro–stock law proclamations tended to focus on the matter-of-fact tenets of progressive agriculture. One beat twenty-eight resident made his case for the law by listing the paltry amounts and inferior quality of meat killed by local farmers. At a local drugstore conversationalists risked eternal damnation for the temporal pleasures of heated argumentation. After several intense discussions threatened to become brawls, the members of the "Stove Club" at Long's Drug Store in Jasper voted to table the topic permanently.[20]

Significantly, the local-option law came on the heels of passage of the 1901 state constitution that severely limited the franchise among blacks and poor whites, the two groups most threatened by stock laws. According to Sheldon Hackney, the coalition of Black Belt planters and Big Mule industrialists established an "elaborate maze through which one had to grope to claim the privilege of becoming an elector." This maze of literacy and property qualifications capped off by a $1.50 poll tax disfranchised almost all Alabama blacks and a significant proportion of poor whites, making it easier for planters and agricultural progressives to carry local elections and close the range.[21]

The stock law controversy also created a bevy of court cases concerning complaints of stray stock and allegations of fraudulent elections. In 1907, the Alabama supreme court overturned a ruling of the Elmore County circuit court that had awarded payment to a farmer for crop damages rendered by a lumber company's cattle. The court ruled in favor of the Clear Creek Lumber Company because the farmer's land lay in an open-range district. In 1905, the court upheld a ruling by the Marengo County circuit court fining J. S. Ryall $125 for his stock's destruction of hay and cotton fields belonging to L. J. Allen. In this case Allen's lands were within a stock law district. In another case the supreme court ruled that stock owners in open-range precincts were liable for damages wrought by their stock in closed-range beats. This Mobile County incident resulted after employees of the Chickasaw Land Company seized and held cattle belonging to cattlemen in an adjacent open-range district.[22]

A few cases challenged the constitutionality of stock laws. The supreme court repeatedly upheld their constitutionality or, more often, simply avoided the issue by declaring rulings on the basis of other case issues. For instance, in 1880 the court reviewed a case from Sumter County circuit court involving damage claims. Six years earlier William Winston had sought a settlement of twenty-five dollars for crops destroyed by Orrin Joiner's cattle. Since both men resided in a stock law beat, the local court ruled in favor of Winston. Joiner appealed the ruling and asked that the supreme court rule on the constitutionality of the legislative action establishing the stock law district. The supreme court instead overturned the lower court's ruling on the basis that the Sumter County court of commissioners had declared the district a closed-range area despite the fact that only a minority of landowners in the beat favored the stock law.[23]

After the institution of the 1903 local option law, suits challenging elections proliferated. The evidence in these cases often revealed the stark divisions in the range controversy and the possibility of balloting shenanigans. In February 1914, the voters of DeKalb County's tenth precinct went to the polls to decide the range question. After a vote showing thirty-nine for the stock law, thirty-eight against, and one mutilated and unascertainable ballot, the county board declared the election a free range victory. The proponents of the stock law promptly won a ruling in their favor from the local probate court after presenting evidence that one voter against the stock law had failed to pay his poll tax six years earlier. When the anti–stock law men appealed, the supreme court reversed the lower court's decision, finding on the same evidence that three men who voted to close the range were non-residents of DeKalb County at the time of the vote. Thus, after nearly a year of turmoil and litigation, the open-range advocates enjoyed a temporary victory.[24]

Railroad companies were often the most conspicuous stock law proponents. Claiming that they paid thousands of dollars annually in restitution for livestock killed by trains, railroad companies sought to avoid losses by promoting stock laws. Over most of the late nineteenth century Alabama law favored farmers, holding railroads liable for killing or injuring stock. An 1852 act of the legislature made railroads absolutely liable for such damages, regardless of the circumstances. Six years later an act altered the earlier law, hinging the railroad's liability on its failure to take prescribed precautions, which included blowing the whistle and slowing when necessary. Two decades later the Alabama supreme court blocked a reversal of this measure when in *Zeigler v. South & North Alabama Railroad Company* it declared unconstitutional an 1877 statute imposing absolute liability on railroads.[25]

Though railroads continued to be responsible for the burden of proof in such cases, their legal position improved significantly over the years. In the late 1870s the Alabama legislature prohibited the "salting" of tracks, an apparently not uncommon practice by stock owners to entice scrub or sick stock to their death and thereby collect payments. Furthermore, the supreme court often reversed lower-court decisions ordering railroads to pay for injured or killed stock. For instance, in 1871 the supreme court overturned a lower court's ruling awarding a Mobile man one hundred dollars for four cows killed by trains on the Mobile & Ohio line. The ruling was based on the stockman's failure to present his claim to an official railroad representative within the time allowed.[26]

By World War I railroads continued to pay thousands of dollars annually for destroyed livestock. An official of the Louisville and Nashville Railroad complained that his company had paid out to Alabama stockmen more than $50,000 in the first six months of 1918 alone. In an effort to reduce the number of livestock deaths on rights of way, the federal government and the State Council of Defense took measures during the war such as passing out pledge cards on which stockmen promised to keep their cattle off railroad tracks. The Council of Defense distributed the pledge cards as part of the "Livestock Conservation Campaign," which was funded by railroad companies. Again, this government-sponsored program drew the ire of some cattlemen. N. B. Cravy of Green Bay (Covington County) found the campaign's requests "unreasonable as this Country hear is lots woods land and a good rang for stock." He argued that "the farmers have got somthing els to do" besides spend their days keeping cattle off rights of way.[27]

The closing of the range slowly removed a major barrier to the growth of midwestern-style cattle raising and ignited social and political upheaval along the way. While stock laws transformed the agricultural structure of the countryside, state and federal government initiatives constructed the bases for experimentation and scientific development that would play an integral role in the transformation of Alabama agriculture in the twentieth century. In 1872, ten years after the Morrill Land-Grant Act had provided federal funds for agricultural and mechanical colleges, Alabama established its white land grant institution at Auburn on the site of a struggling Methodist college. Inheriting four professors and 103 students from the old college, the new Agricultural and Mechanical College of Alabama continued for some years to concentrate on a classical liberal arts education. Nevertheless, the institution's new mission of agricultural and mechanical education and the "New South" philosophy of president Isaac Tichenor inevitably led to increased attention to farming and agricultural research.[28]

Tichenor's New South creed called for a new socioeconomic structure for the South, one based on industrialization and scientific, diversified agriculture. Tichenor realized that the interests of Alabama farmers lay in the production of cotton but that, with the development of cheaper, less time-consuming cotton-raising methods, Alabama farmers could devote more energy toward diversification. Consequently, Tichenor directed the college in purchasing abandoned farmland near the campus for use as demonstration farms. These lands would eventually form the nucleus of the Alabama Agricultural Experiment Station.[29]

These early demonstration farms tested the effects of fertilizer on the state's popular crops. During the 1870s and 1880s, growing numbers of Alabama farmers began to utilize commercial fertilizers. As a result of the questions surrounding the new fertilizers, agricultural organizations such as the Grange began to push for state inspection laws. In 1883, the Alabama legislature passed such a law, Hawkins' Bill, which also formally created an agricultural experiment station.[30]

The Hatch Act of 1887, the product of five years of effort by agricultural educators and congressmen, provided much-needed federal funds for state-level agricultural research. Though most early research focused on cotton, corn, and other row crops, some programs dealt with the livestock industry. In 1888, the experiment station ventured into dairy research when an Opelika dairyman, Isaac Ross, agreed to furnish the station with a small herd of Jersey cattle for a three-year period. Ross was to receive an annual salary of $1,700 for overseeing the operation and was allowed to market the butter produced on the farm after purchasing it from the station. In 1890, citing the inefficiency of the project, the experiment station's director convinced the college trustees to break the agreement with Ross, effectively ending the project.[31]

In the meantime a branch of the experiment station, the Canebrake Agricultural Experiment Station, began research on livestock feeding. In the early 1880s, the Alabama Agricultural Society and experiment station director James Newman, realizing the diversity of Alabama soils and the unrepresentative nature of Auburn's soils, urged the legislature to create a substation in the Black Belt, the most politically powerful and agriculturally profitable section of the state. In 1885, the legislature agreed to the proposal and authorized the purchase of forty acres near Uniontown in Perry County. The Canebrake Station began receiving federal subsidies after

Two symbols of progressive agriculture in the South: Jersey cattle graze in front of Comer Hall at the Alabama Polytechnic Institute (Auburn University). (Courtesy of ACES Photo Collection, Auburn University Archives)

passage of the Hatch Act and launched a series of experiments involving various phases of agriculture. Despite the dominance of experiments with cotton, corn, wheat, fruit, and vegetables, the Canebrake Station's interest in livestock reflected the inheritance of antebellum concerns of Black Belt planters as embodied in the Alabama State Agricultural Society. In the late 1880s and 1890s, researchers at Uniontown published bulletins on forage crops, pasture grasses, and cattle and swine feeding. The substation even brought some of the earliest thoroughbred beef cattle into the region for growth comparisons with native stock.[32]

The main experiment station at Auburn took an important step in 1899 with the decision to reinstitute its livestock research. In the winter of 1899–1900, the station purchased a number of purebred beef calves from Starkville, Mississippi. Aided by a U.S. Department of Agriculture grant of $5,000 in 1905, the station's combined beef and dairy herd had grown to fifty steers and twenty brood cows by 1906.[33]

In addition to the research conducted at Auburn and Uniontown, experiment station personnel soon began cooperating with successful stockmen to establish experimental herds. Though not demonstration herds as later defined by cooperative extension, these operations often influenced neighbors to start raising beef cattle. Beginning in 1906, several prominent cattlemen agreed to allow representatives of the experiment station to observe and record their herds' progress from year to year. In the Tennessee Valley, Lauderdale County planter and stockman J. S. Kernachan, one of the state's earliest breeders of registered Aberdeen-Angus cattle, served as an experimental stock raiser. Black Belt cattlemen E. F. Allison and F. I. Derby, both of Sumter County, invited station workers to chart the growth of their mixed-blood stock. As a result of these studies, the Alabama Agricultural Experiment Station published some of its first bulletins on raising beef cattle, and switching to beef cattle production became a viable option in the fight for diversification.[34]

In addition to the land grant institution and the experiment station, two other government-directed efforts helped pave the way for diversification and the establishment of the modern cattle industry. The cattle tick eradication program and cooperative extension service both grew out of responses to potential agricultural disasters. The cattle tick (or fever tick) plagued stockmen for years and posed a direct threat to cattle raising. The cotton boll weevil invasion, which spawned the extension service, indirectly bolstered the livestock industry by forcing diversification efforts on desperate farmers.

The Texas fever tick, unlike the boll weevil, was not a new arrival in the twentieth century. The ticks, brought to the United States on Spanish cattle and spread throughout much of the country by the trail drives from Texas into the Midwest, transported microparasites from sick cattle to healthy ones. Northern winters usually killed the ticks, but the warm, humid South provided a year-round home for the pests. The parasite destroys red corpuscles in the cow's blood that are required to break up waste material. Within a week to ten days infected cattle can show signs of high fever, appetite loss, coughing, weakness, flesh loss, accelerated breathing, or any number of other symptoms. Early methods of eradication included cleaning cattle with oils, dipping them in arsenical or coal tar solutions, or simply picking off the ticks by hand.[35]

Federal authorities initiated the organized fight against the cattle tick in 1884 when the USDA created the Bureau of Animal Industry. In the South, North Carolina first accepted the challenge of eradicating the pest. Between 1899 and 1906, state and federal authorities collaborated to free a dozen counties in that state from the federal quarantine that covered most of the region. In 1906 Congress empowered the secretary of agriculture to initiate a plan of state and federal cooperation in the eradication effort. The federal Department of Agriculture agreed to supply trained supervisors, or "tick inspectors," to participating counties.[36]

The tick eradication program promptly commenced in two Tennessee Valley counties, Madison and Limestone, under the rules of a state livestock sanitary law. The proponents of the eradication program suffered a setback in 1907, however, when the state legislature, in an attempt to avoid difficulties and controversy, declared the livestock sanitary law inoperative in those counties with more than half their lands in open-range districts. The determined state veterinarian, Dr. C. A. Cary, advised officials to carry on the project in those counties dominated by stock law beats, namely the Black Belt and the Tennessee Valley.[37]

Charles Allen Cary was the most influential noncattleman in the modern Alabama cattle industry and perhaps the single most important figure of his era in regard to southern livestock raisers. Born and raised on an Iowa farm, Cary completed his training in veterinary medicine at Iowa State College in 1887. After a few years of private practice and teaching, he arrived at Auburn in January 1892, where, except for a few months spent in Germany doing postgraduate work, he remained for more than forty years. While serving as professor of physiology and veterinary science and as experiment station veterinarian, Cary received an appointment as Alabama's first state veterinarian in 1905. Two years later he was named dean of the fledgling College of Veterinary Medicine at the Alabama Polytechnic Institute, the first school of its kind in the South. He held that position until his death in 1935.[38]

In addition to being a progressive-minded civic leader, Cary was a "major southern participant" in the realization of the nineteenth-century's microbiological revolution, characterized by "disease control campaigns and other forms of mass actions . . . designed to protect large livestock populations."[39] The red-haired, mustached veterinarian wielded political influence far beyond that which his five-foot seven-inch, 150-pound frame might have suggested. Cary lobbied for and procured state legislation requiring meat and milk inspection and aimed at the eradication of bovine tuberculosis. He was also the architect behind Alabama's tick eradication laws.[40]

After the setback of 1907, Cary continued to lobby for a statewide, mandatory tick eradication law. In the meantime, as secretary of the State Live Stock Sanitary Board, he helped direct the program in stock law counties. In 1908, state and federal authorities instituted eradication programs in fifteen counties, primarily in the Black Belt and Tennessee Valley. Without a statewide eradication law, however, many officials remained pessimistic. In June 1911, the *Montgomery Advertiser* ran the headline "Tick Eradication in Alabama Fails." Fortunes changed for the progressive agriculturists, though, with the election of sympathetic governor Charles Henderson in 1914.[41]

In January 1915, a combined meeting of the Montgomery Cattlemen's Club, Southern Cattlemen's Association, Alabama Live Stock Association, and Blackbelt Livestock Association launched a "State Campaign for the Eradication of the Cattle Tick." Again, C. A. Cary headed the committee to draft a bill. Ten representatives visited Governor-elect Henderson at his home in Troy, where they emphasized the financial importance of the program. According to the group the quarantined sections of the South annually lost 20 percent of the value of their cattle. (Two years earlier cattlemen in newly unquarantined counties had reported a market increase of $7.70 per cow.) Not by chance, this meeting of influential central Alabama agriculturists coincided with the arrival of the boll weevil, another problem that required definitive action.[42]

The cattlemen's meeting with the governor must have been productive. In 1915 Governor Henderson dubbed the week beginning April 26 "Tick Eradication Week in Alabama." By the end of 1915, several Tennessee Valley and Black Belt counties had been freed of the federal quarantine. Furthermore, "tick inspectors" successfully initiated eradication efforts in a dozen infected counties in the northern hills and southern piney woods, such as Walker, Marshall, Baldwin, and Escambia.[43]

These tick inspectors used the dipping vat in their eradication efforts. The vats, usually built at public expense except for those used on some of the state's largest farms, were generally six- to seven-foot-deep underground concrete tanks with a narrow chute on one end and a holding or dripping pen on the other. The cow entered the vat through the chute and plunged into six feet of a chemical solution (generally an arsenical solution). After swimming across the twelve- or fifteen-foot-wide vat, the animal climbed out the other side by means of steps on an incline. After the animals had remained in the dripping pen for the necessary length of time, the owner could drive or haul them back home. Each cow had to be dipped once every two weeks during the fall and spring dipping periods and, upon being dipped, received a painted spot on its side indicating its compliance. In most tick-infested areas, all cattle, not just those noticeably infected, were required to be dipped at the nearest vat.[44]

As has so often been the case with domestic social and economic change, the exigencies of war provided the final inducement for the transition at hand. At the time of the United States' entrance into the Great War, several mountainous and piney woods counties remained under quarantine. In November 1917, as a "war measure," the State Live Stock Sanitary Board declared, as of January 1, 1918, a ban on all intrastate or interstate shipping of tick-infected cattle. The state assembly followed this measure with a 1919 law establishing a statewide eradication program. The new law mandated the program for all counties at least partially infested with the tick. The bill required county commissioners or the board of revenue to build the inspector-specified number of dipping vats and supply the necessary chemicals. The law forced livestock owners to dip their cattle every two weeks at the appointed place and time and imposed a fine of ten to five hundred dollars on those failing to comply. Dr. Cary had finally achieved his goal.[45]

The struggle did not end there, however. Compelled by the visions of the New South and the ideals of agricultural progressivism and directed by the inherited experience of midwestern agriculture, Cary little understood the southern open-range tradition and the people who practiced it. The fiery, persistent veterinarian embodied the justly inspired but often narrow and conformist notions of the progressive spirit. Many farmers responded to the interventionist program with the hostility and derision exhibited in the range controversy. The interventionist spirit of the eradication program was not shrouded in the cloak of democracy, however, and consequently the reactions to the law's demands proved violent and hysterical.[46]

As historian William A. Link has observed, many "traditionalists" in the southern countryside "opposed strong centralized power and resisted the intrusion of large, impersonal forces into matters heretofore under community control." The clash of these traditionalists with the more powerful reformers, who "saw solutions through the expansion of coercive state intervention," comprised "a central paradox of early twentieth-century reform." Like the range controversy, cattle-dipping protests reflected class antagonisms. The poor often felt removed and alienated from the centers of power and distrusted the "book learning" methods of government personnel. One Coosa County man remembers referring to those who opposed mandatory dipping as "seconds," because of their lower-class status in the community. These struggles further reflected the clash of the traditional, open-range system with the blossoming midwestern-style herding operation. Because most native cattle had developed an immunity to the tick-borne disease, its presence often went unseen until someone imported purebred British stock into a locale. One Covington County cattleman recalls this as an important point of contention among local livestock raisers. After a local farmer lost some newly acquired Devon cattle to tick fever, he and other farmers began to encourage dipping, though the poorer owners of native cattle opposed the program as superfluous.[47]

Opponents of mandatory cattle dipping protested in various ways. One DeKalb County newspaper editor linked the tick eradication law with the stock law and claimed that both threatened livestock raising. He also stressed the common theme that government officials overestimated the damage done by the cattle tick.[48]

The most common form of protest was refusal to bring cattle to the dipping vat. Several farmers and stock raisers were fined or arrested for failure to dip their cattle. Most of these resided in the backcountry areas of the mountains or piney woods. In July 1919, eleven Geneva County tick law violators received fines of one hundred dollars each. That same month local authorities arrested twenty-five farmers in Etowah County for refusing to dip their cattle. Dozens of others around Alabama suffered similar outcomes for failure to abide by the tick law.[49]

Other disgruntled farmers took action to remedy the perceived injustice of government intervention. As a last futile attempt to avoid this intervention, several people dynamited local dipping vats. An April 1919 issue of the *Cullman Democrat* reported that three of the county's 150 vats had been blown up during the previous week. Local authorities used bloodhounds to track the culprits, and one man was arrested. A week later, after the Sulphur Springs vat had been repaired, guards shot at two men carrying dynamite near the vat. In July of the same year dynamiters destroyed a vat twelve miles south of Gadsden. An angry local editor proclaimed the men had "accomplished nothing except to stamp themselves as outlaws and enemies of orderly government." In the summer of 1919, the dynamiting of vats became such an epidemic that Governor Kilby offered a two-hundred-dollar reward for each conviction of a vat dynamiter. Such resistance reflected a pattern that developed throughout the South.[50]

The tick eradication law also precipitated legal battles. One Mobile County case affirmed the notion that swampy and open-range conditions were no excuse for failure to dip cattle. In at least one instance C. A. Cary took direct legal action to force compliance with the tick law. In 1927, the state veterinarian brought suit against Clarke County, one of the final three counties under federal quarantine. Cary declared that the county needed to spend several thousand dollars to improve its eradication program, which would include construction of fifteen new vats. After the Clarke County circuit court ruled against Cary, the state supreme court reversed the decision, forcing the county to carry through with the program.[51]

Cary's legal victory propelled his cause to victory after two decades of struggle. On December 1, 1929, the federal Bureau of Animal Industry lifted the quarantine from Clarke County, thus freeing the entire state of Alabama from tick-induced federal restrictions. The forces of progressive reform had proved victorious in this controversial but now largely forgotten struggle. The successful suppression of the traditional agrarian enmity toward government intrusion in local life coincided with the onset of an economic depression that would further test the weakened resolve for independence and freedom from intervention among the impoverished descendants of the frontier. Both these forces, progressivism and depression, paved the way for the widespread acceptance and appreciation of New Deal activism.[52]

Another agricultural product of the Progressive Era was the cooperative extension service. The cotton boll weevil provided the primary impetus for this development. The pesky insect, according to historian Pete Daniel, "upset the traditional culture enough so that southern farmers were willing to look up from their almanacs and listen to agricultural experts." The Mexican boll weevil, which jumped the Rio Grande into the Texas cotton fields in the 1890s, steadily made its way northward and eastward into the heart of the cotton South. In 1902, the USDA took its first steps toward federal demonstration when it hired renowned agriculturist Seaman A. Knapp to show Texas farmers how to combat the damaging weevil through improved cultivation techniques.[53]

Entomologist Warren E. Hinds first found weevils on Alabama soil on September 3, 1910, in western Mobile County. By the end of 1911, the pests had infested all or parts of a dozen southwestern Alabama counties. In that same year the state assembly voted to provide funds for the creation of an Agricultural Extension Department under the direction of the experiment station. Faced with the necessity of educating thousands of cotton farmers, the new extension service supplemented, and eventually replaced, the work of the old Farmers' Institutes, which had been operated from Auburn and Tuskegee for two decades.[54]

By 1914, when the Smith-Lever Act provided federal funds for extension work, the boll weevil had overtaken more than half the state's territory. County extension agents urged farmers to diversify, to substitute livestock and fruit and vegetable crops for portions of their cotton acreage. Nevertheless, the methods of diversification were most often costly, and the majority of Alabama's farm families lacked the financial resources or knowledge to carry out the extension service's plan. Consequently, as Pete Daniel observes, "the Extension Service grew into a bureaucracy that increasingly served commercial farmers who controlled agents on the local level."[55]

In addition to spawning the development of the cooperative extension service, the invasion of the boll weevil exercised an immediate impact on farming in certain areas of the state. Extension agents in the Wiregrass region promoted the raising of peanuts and swine. Many farmers of that region converted all or part of their cotton lands to peanut cultivation; the remains of the peanut harvest provided an excellent source of food for fattening hogs for the market. The establishment of a packing house in Andalusia was both a response to the rapid growth of hog raising in the Wiregrass region and an impetus for further growth. In his work on the American livestock industry, Rudolf Alexander Clemen suggests that the appearance of the boll weevil marked the beginning of the modern southern cattle business. This modern cattle business experienced its most significant growth in the Black Belt, where many planters and farmers invested in beef and dairy cattle to offset the effects of the weevil's destruction.[56]

Before the arrival of the boll weevil, the tremendous fertility of the black, limestone soil in the Black Belt made the region the most agriculturally productive area of Alabama. The heart of the Black Belt, Dallas County, annually led the state in cotton production, easily surpassing the nearest competitors. For a century the Black Belt served as the model of the cotton-plantation South in a cotton state. Its population boasted many of the state's wealthiest, best-educated, and most-influential white citizens, who wielded political power disproportionate to their relatively sparse numbers.[57]

Long before the arrival of the boll weevil, private and governmental forces were establishing a foundation on which to build a modern, midwestern-style cattle industry. The region's political clout weighed heavily in the location of the agricultural experiment station's first branch near Uniontown. By the end of the nineteenth century, area planters were familiar with the Canebrake Station's experiments with pasture grasses (alfalfa was first raised on Alabama soil at the station), forage crops, and purebred beef cattle. In addition, the Black Belt counties were among the first to eradicate the cattle tick with the help of state and federal government authorities. Furthermore, some of the region's most successful stock raisers served as demonstration farmers for experiment station research.[58]

Private organizational and individual initiative also played a key role in the development of the Black Belt cattle industry. This development proceeded according to the midwestern livestock raising model. While Alabama farmers planted more and more cotton and government programs slowly laid the foundation for future diversification efforts, beef cattle raising had evolved into an efficient, scientific industry concentrated in the Midwest. The introduction of barbed wire, the demand for marbled meat, and the widespread dispersion of thoroughbred beef breeds throughout the northern and midwestern states precipitated the development of a system of cattle operations in Missouri, Iowa, Kansas, and other states. This system, a product of progressive agriculture and the northern farming and livestock raising traditions, achieved a dominant position in the American livestock industry. The success of this system encouraged other regions to emulate its structure and methods. This the Black Belt accomplished through the gradual transference of agricultural methods and the response to the boll weevil emergency.

The nongovernmental factors influencing the growth of the Black Belt cattle industry were numerous. As we have seen, the region's tradition of livestock improvement and diversification stretched back to the antebellum era. After the Civil War some of the state's largest cattle raisers operated Black Belt plantations. Because legislative mandates and local boards closed the range in the bulk of the region before 1900, stock owners lacked only purebred cattle and a knowledge of pasture grasses and forage crops in order to establish a cattle industry profitable enough to supplement or supplant cotton raising. Many of the region's planters also possessed the key elements that prevented small farmers and tenants from diversifying into livestock raising: large landholdings and capital. Hazel Stickney, in her study of the transformation from cotton planting to cattle raising in the Black Belt, observes that "the initial bigness of landholdings probably facilitated the change to beef production more than any other single circumstance." Even a gradual program of diversification could cost thousands of dollars at the beginning, including the expenses of purchasing thoroughbred cattle and converting cotton acres to pasture or forage lands. Furthermore, most of the region's earliest cattlemen were also involved in businesses and professions other than planting. These individuals were in the best financial position to establish large cattle herding operations.[59]

One example of the planter–livestock raiser was Will Howard Smith. One of central Alabama's most influential cattlemen in his time, Smith grew up four miles east of Prattville on the plantation of his father, McQueen Smith. After graduating from Iowa State College in 1906 and having observed the successful cattle operations of the Midwest, he returned to Autauga County where he helped his father diversify their operation and enlarge and improve their cattle herd. By World War I, McQueen

Smith and Sons owned about 800 head of cattle and four purebred, registered bulls, in addition to their lands in cotton and other crops. Smith, like many other prominent planter-cattlemen, branched out into the corporate world; he eventually served on the boards of directors of Alabama Power Company and the Birmingham branch of the Federal Reserve Bank.[60]

Stonewall McConnico was the eldest son of Wilcox County planter William Washington McConnico. He served for a time as county sheriff and continued to operate the family plantation; McConnico also developed one of the early Angus herds in the state.[61] Two other Black Belt scions, Robert Goldthwaite Arrington and Robert James Goode, Jr., reflect the social standing of many pioneer Black Belt cattlemen. The son of a distinguished lawyer and grandson of a planter-lawyer, Arrington attended the University of the South and the University of Alabama law school. Admitted to the bar in 1901, he rose to a prominent position in civic and legal circles in Montgomery, as well as among the participants in the Black Belt's infant cattle industry. Arrington displayed stock from his purebred Hereford herd at Montgomery's first exclusively livestock show in 1915 and remained an influential figure in Alabama's cattle industry throughout the pre–World War II era. Goode, of Gastonburg in Wilcox County, graduated from Marion Institute and briefly attended the University of Pennsylvania. He returned home to the family plantation before World War I, where he became a planter, a booster for the Hereford breed, a local insurance businessman, and a member of the state legislature.[62]

Other individuals traveled more circuitous routes to establish modern cattle operations. John Eugene Dunaway, a Wilcox County native and Marion Institute graduate, started his career as a bookkeeper in Dallas County. After marrying into a prominent Orrville family in the 1890s, he established the J. E. Dunaway Mercantile Company and the Orrville Bank and Trust before being elected to the state assembly. Dunaway eventually acquired some 6,000 acres, West Dallas Farms, which he operated with Joe Lambert, a successful cattle buyer who shipped between 400 and 500 railroad carloads of stock each year. In addition to having a herd of 200 registered Herefords and many other cattle, West Dallas Farms produced thousands of bales of cotton.[63]

A look at a few other cattlemen active in breed associations or in the Alabama Live Stock Association reveals the cattle industry's inheritance from prominent agricultural and business families. Lucien Powell Burns, a Dallas County Hereford breeder, was born into a prominent Selma family in 1889. Educated at Howard College (now Samford University) and Washington and Lee University, he operated a wholesale grocery company in Selma, served as director of the chamber of commerce, and was elected president of the city council. Montgomery Hereford breeder H. S. Houghton attended Alabama Polytechnic Institute and earned a law degree from the University of Alabama law school. In addition to his legal career and interest in cattle raising Houghton served as director of the Alabama National Bank and as treasurer of the Alabama Farm Bureau Cotton Association. Another prominent Montgomery man, Gaston J. Greil, owned a herd of Angus cattle, though cattle raising was far from being his primary occupation. A graduate of Alabama Polytechnic Institute and Columbia University, the pediatrician, whose wife had attended Vassar College, served a term as president of the Montgomery County Medical Association and was an officer in the Alabama National Guard.[64]

These men demonstrate an important characteristic of the early Alabama leaders in the modern cattle industry: cattle raising provided a relatively insignificant portion of their incomes. One particular biography in the 1929 edition of the *Alabama Blue Book and Social Register* strikingly reveals the irony of this pre-Depression diversification movement. William Edward Elsberry, Jr., manager of the Burton Transfer Company, was a founding member and secretary of the Montgomery Livestock Association before World War I and was also active in the country club and Rotarians; however, he listed farming only as his hobby. The success of the pre–World War I drive for diversification was realized on the farms of the people who least needed it and least depended on the land for a livelihood. This modern cattle industry was not for the light of purse. In a time when a Lowndes County planter could pay a Kentucky stockman $1,200 for a Hereford bull, Alabama's per capita rural wealth ranked last in the United States, well below half the price of that bull. Nothing an extension agent might say could change the economic reality. For many of these new "cattlemen," the raising of livestock represented an alteration or extension of the traditional, absentee planter role, not the pinnacle of agricultural progressivism.[65]

In addition to the advantage enjoyed by some of the region's long-time inhabitants, other factors proved crucial to the early prominence of Black

Belt cattle raising. One key to understanding the pre–World War I growth in cattle raising is the occurrence of direct technical and cultural transferal. Several midwesterners purchased Black Belt farms in the first two decades of the twentieth century and established modern cattle operations modeled on those of their native region. These immigrants liked the mild winters, good rainfall, natural pasture soils, and level to rolling terrain of the region. Furthermore, the improvements in technology and veterinary science rendered the hot, dry summers and disease-breeding humidity less foreboding. These men became influential in the promotion of livestock raising in Alabama and provided an example of diversification on the state's own lands. The other midwestern influence came in the persons of breed association agents. These representatives from the three big breed associations—Shorthorn, Hereford, and Aberdeen-Angus—infiltrated the Southeast in ever greater numbers before the end of World War I.

Between 1903 and 1905, a colony of about one dozen livestock-raising families from the Bluegrass region of Kentucky settled in Montgomery County east of the capital city. These cattlemen were among the first to import purebred beef cattle into central Alabama. Another native of Kentucky, Morton C. Crabb, raised one of the state's first registered Shorthorn herds on his Gallion farm in Hale County. In 1915, the Farm and Immigration Section of the *Montgomery Advertiser* ran an article on C. I. Stoner, an Indiana man who maintained a large Angus herd on his farm near Demopolis. Dallas County cattleman J. E. Dunaway even recruited a Kentucky stockman to manage his West Dallas Farms herd. The immigration of midwesterners reached a level sufficiently significant to prompt the following *Montgomery Advertiser* headline: "Will All Our Black Belt Lime Lands Go to Central West Farmers?"[66]

Several midwestern cattlemen found their way to the Southeast during the boom days of the First World War. In 1918, two Ohio natives, Charles W. Rittenour and T. B. Irving, purchased a 2,800-acre plantation just outside Montgomery. After developing a successful cattle-raising operation, Rittenour became one of central Alabama's most active and influential cattlemen. In her study of the Black Belt cattle industry, Hazel Stickney chronicles two Dallas County families with roots in the Midwest. Both the Wallaces and Caleys settled in the Black Belt in 1918, and both families established cattle farms. The patriarch of the Wallace family had become familiar with the region on his trips there from Kentucky to purchase cattle.[67]

Despite these scattered achievements, the Black Belt cattle industry crawled along in infancy before the arrival of a serious threat to the profitability of cotton. This threat, of course, was the boll weevil. The weevil entered the western fringe of the Black Belt as early as 1911, and by 1915 the entire region had been infested. Within a decade researchers and extension agents would provide Alabama's farmers with insecticides and early-maturing cotton, drastically reducing the level of weevil destruction. However, the Black Belt soils proved inhospitable to early cotton. The "cold clays" of the dark earth prevented early spring planting, thereby exposing the region's slow-maturing crops to the hungry hordes of summer weevils.[68]

Another development aided the weevil in hindering cotton production while giving direction to diversification efforts. For several years many Black Belt planters had struggled against the persistent Johnson grass. First planted by the region's agriculturists in the late nineteenth century, the perennial tall-stemmed plant had gradually spread its way across western and central Alabama to the chagrin of planters, who attempted to keep Johnson grass from choking out their cotton crops. At the same time, planters with an eye for diversification had grown to appreciate the uses of Johnson grass for hay and pasture. Planters around Marion Junction began growing hay commercially in the 1880s, and by 1910 the town shipped hay all over the country at the rate of one carload each day. The presence of Johnson grass and other perennial grasses, bolstered by the Canebrake Agricultural Experiment Station's pasture research, made cattle raising a logical alternative in the face of the boll weevil attack. Hay growing continued to gain such importance among Black Belt planters that after World War I a group of Hale County farmers formed the Hay Growers Association, which marketed Johnson grass hay for $3.50 per ton.[69]

The result of the boll weevil's arrival and the availability of avenues of diversification was a tremendous decline in cotton cultivation. This development reached particularly significant levels in the Black Belt (Table 2, appendix). Between 1910 and 1920, the acreage devoted to cotton in the ten core Black Belt counties plummeted by more than 50 percent from more than 1.1 million to 531,000, while the rest of the state experienced only a 20-percent drop. Only Dallas and Sumter counties experienced declines of less than 50 percent, and even in these counties the 1920 crop represented less than two-thirds the size of the 1910 crop. Montgomery County underwent the most drastic reduction: the 48,625 cotton acres of 1920 revealed an almost 70-percent decrease from the number a decade earlier. Over the same period the Black Belt acres harvested for hay more than doubled to 127,000. Only three counties realized increases of less than 100 percent, and both Bullock and Perry county farmers quadrupled their hay acreage.[70]

The decade before World War I witnessed a frenzy of agricultural activity as planters and farmers prepared for and reacted to the boll weevil's arrival. Newspapers began to report stories of diversification, hailing those responsible for this progress as heroes worthy of emulation. Government representatives and influential farmers cooperated to carry out programs aimed at discovering alternatives to cotton. And breed association representatives began to circulate throughout the Southeast, encouraging Alabamians to purchase animals from large midwestern farmers.

In 1906, the Alabama Department of Agriculture and Industries reported the story of the state's first large, modern cattle ranch in Hale County. W. W. Murphy's ranch, comprising three former cotton plantations, held some 1,100 head of cattle, many of them registered Angus, Hereford, and Shorthorn. Murphy raised Johnson and Bermuda grass and vetch and shipped his cattle to northern markets. Hazel Stickney notes that Sumter, the first Black Belt county hit by the boll weevil, possessed some of the state's earliest purebred cattle operations. In 1910, T. M. Tartt, a Livingston bank president, and three other men brought herds of registered Shorthorns and Herefords from Texas. Between 1910 and 1915, several other area planters and businessmen got involved in the cattle business as well.[71]

The *Montgomery Advertiser,* like many other southern newspapers, took up the cause of diversification and livestock promotion. In the fall of 1910, a reporter for the newspaper, relating the story of Dallas County stockman A. D. Summers, wrote, "It is manifest that cattle can be grown and sold in the grass fields of West Dallas at a very satisfactory profit." Five years later the *Advertiser* observed that a "noted western stockman" had called the South a "great livestock region" and had urged southern bankers to make loans for livestock purchases. Another article applauded the diversification efforts in Dallas County where farmers planned to convert 25,000 cotton acres to pasture and forage fields in 1916.[72]

The rebirth of diversification efforts also accomplished the rejuvenation of the livestock show and fair. After the initial success of the livestock show at antebellum state fairs, the intensification of cotton planting after the Civil War diverted attention away from livestock exhibits at state and local fairs into the twentieth century. The boll weevil threat provided the impetus for a livestock show at Marion Junction in October 1910. Possibly the first exclusively livestock show in Alabama, the Marion Junction event featured displays of area Herefords and Jerseys and attracted about 2,000 visitors.[73]

Five years after the inaugural event at Marion Junction, the Montgomery Livestock Association sponsored the first annual Montgomery Live Stock Show, a larger fair representing prominent, purebred cattlemen throughout central Alabama. The three-day November event offered awards in five cattle categories: Angus, Hereford, Shorthorn, Jersey, and Holstein. Exhibitors included western Alabama cattlemen Oscar Cobb and F. I. Derby of York, as well as several influential "hobby" farmers. Among these Montgomerians were manufacturer and Virginia-native Ralph D. Quisenberry, lawyer and manager of the Alabama State Fair Simon Roswald, and attorneys Robert G. Arrington and H. S. Houghton. At the second show in 1916, representatives of seventeen counties formed the Alabama Live Stock Show Association for the coordination of the growing event.[74]

The popularity of these livestock shows reflected the increasing number of Alabama farmers and stockmen raising purebred and registered cattle. The chief indicator of a modern cattle farm, the purebred bull or cow was a relative newcomer on the Alabama scene in the early twentieth century. Planters in the Black Belt and Mobile areas had purchased a few thoroughbred cattle during the late antebellum period, only four of which were registered with their respective breed associations. The vast majority of these antebellum thoroughbred cattle were used as dual-purpose animals, serving both dairy and beef purposes. The Civil War damaged this phase of the cattle industry. For three decades after the war, those few agriculturists acquiring thoroughbred cattle turned their attention to dairy production. Consequently, Jerseys dominated the small field of purebred stock. By 1890, Alabama counted fewer than 2,800 purebred cattle.[75]

Although subsequent censuses stopped enumerating purebred animals, we may find some indication of the growing numbers of purebred cattle and their breeders through the use of breed association herd books. In 1901, W. C. Swoope of Courtland registered the first Alabama-born Hereford bull. The next year Lauderdale County planter J. S. Kernachan logged Alabama's first Angus bull. By 1907, Kernachan remained the only Alabama member of the Aberdeen-Angus Association, and the *American Short-Horn Herd Book* listed only four breeders in the state. As we have seen, the purchasing and raising of registered and purebred cattle increased significantly after about 1910. Interestingly, the two cattlemen noted above lived in the Tennessee Valley; nevertheless, Black Belt planters and stockmen surpassed the rest of the state's cattlemen in the registration of thoroughbred beef breeds. In a 1916 edition of the *American Short-Horn Herd Book*, the region claimed half the state's registering cattlemen and over half the cattle registered. Of the two dozen Alabamians registering Herefords in 1915 and 1916, thirteen lived in the Black Belt. A list of members of the Aberdeen-Angus Association reveals a definite concentration in the region; the Black Belt supplied eleven of sixteen Alabama members.[76]

Despite the overwhelming dominance of the Black Belt in the early expansion of cattle raising, agriculturists all around the state took up the practice of livestock raising. By 1917, purebred beef breeders could be found from Fort Payne to Mobile and from Tuscumbia to Lockhart. Outside central Alabama, the Tennessee Valley enjoyed the most success in the march toward modern cattle raising. The large plantations provided the land and capital necessary for some prominent planters to take up cattle raising. By the beginning of World War I, the Huntsville area contained five registered Hereford breeders, and Florence stock owner J. S. Kernachan was one of Alabama's most prominent Angus breeders of the first quarter of the twentieth century. The quad-cities area of northwestern Alabama was home to nine registered Shorthorn breeders by 1916.[77]

The growing interest in purebred cattle appeared promising to midwestern breed associations, which promptly sent representatives into the Southeast. These agents, in turn, recruited potential breeders among the circles of prominent farmers. In late 1915, the *Montgomery Advertiser* reported the efforts of two Hereford Association representatives. F. D. Hengst, southern representative of the *Hereford Journal*, and Hereford Association secretary R. J. Kinzer cooperated with two of the state's largest

Cattle in a northern Alabama corral, date unknown. (Courtesy of Alabama Cattlemen's Association)

breeders, N. J. Bell and J. E. Dunaway, to find other willing cattlemen. E. R. Silliman, a southern representative of the Shorthorn Association, cooperated with Oscar Cobb and F. I. Derby in the "missionary effort." He and other breed agents also regularly attended and helped sponsor livestock shows to promote their associations.[78]

The gradual expansion of the breed associations into Alabama reflected the growth and limitations of the infant modern cattle industry in the state. Through a combination of governmental intervention, private initiative, and reactions to nature's eccentricities, Alabamians laid the foundation for a modern, midwestern-style cattle industry. The programs emanating from the land grant college system encouraged diversification and provided scientific information to farmers around the state. The interventionist spirit of the Progressive Age continued to close the range and forcibly rid the countryside of the cattle tick. The boll weevil threatened the very lifeblood of

Alabama agriculture, encouraging diversification and livestock raising. The small, traditional herder found freedom and prospects shrinking along with the open range, while the planter-entrepreneur-cattleman at the forefront of the rising cattle business inherited methods and ideals from the planter tradition.

This break in the tradition of cattle raising was clearly under way in the early twentieth century; nevertheless, it awaited a more clearly defined manifestation that could come only with improved economic opportunities and an intensification of diversification efforts. The domestic frenzy surrounding World War I would supply those opportunities and efforts. The infrastructure was in place. The war years would present many Alabama farmers with the incentives and opportunities to make cattle raising a profitable enterprise and would offer the state its first significant look at the future of cattle raising.

Chapter 4

The Midwestern Model Meets the South

The two and a half decades between the United States' entry into World War I and its entry into World War II witnessed further adoption of the midwestern style of cattle raising by Alabama agriculturists and a rapid decline of traditional open-range herding practices. The exigencies of wartime food production, the influence of extension agents and agricultural researchers, and federal programs restricting cotton production and encouraging diversification all contributed to the increasing popularity of commercial cattle raising. Because of differences in climate, demography, and tradition, however, midwestern-style cattle raising with purebred British breeds, barbed wire, and stock barns assumed a distinctively southern flavor. Southern cattlemen, found mostly in the plantation districts of the Black Belt and Tennessee Valley, usually continued to plant cotton alongside fields of hay and pasture grasses. Mild winters demanded less attention and care from southern cattlemen. And, just as their enslaved ancestors had done generations before, black farmers, many of them cut loose from the cotton culture, continued to supply the labor on the new cattle plantations.

Ironically, though, the commercial cattle-raising method whose continuity was evident in the early years of the twentieth century bore little resemblance to traditional southern herding, reflecting instead the systematic agribusiness qualities of the cotton plantation. The descendants of planters, not of piney woods herders, would lead the way in the pre–World War II cattle industry. Though pockets of open-range herding would survive until the eve of the Second World War, the ideological and agricultural heirs of Noah B. Cloud and Isaac Croom would shape the cattle industry to the agribusiness dictates of scientific, progressive agriculture and efficiency of production.

Wars throughout history have often had the effect of quickening domestic change or of bringing about swift realization of existing revolutions. This trend is evident even in something so seemingly far removed from the field of battle as the cattle industry. Just as the Civil War had focused intense pressure on a declining open-range herding culture and industry, the First World War supplied governmental and patriotic support for the infant commercial cattle industry. Higher cattle prices, improved marketing opportunities, extension service assistance, and breeder cooperation built on prewar developments to increase farmer interest in cattle raising. Urban growth and wartime demand for dairy products also spurred the growth of dairy farming operations in the countrysides around Alabama cities. Furthermore, the growth of cattle raising played a role in the complex of events surrounding the Black Exodus of World War I.

Despite the sharp postwar downturn in cattle prices, wartime expansion solidified the permanent existence of a substantial cattle industry in Alabama's Black Belt. Though farmers and businessmen slowly took up the practice of cattle raising in all parts of the state and farmers from Robertsdale to Stevenson continued to raise cattle for milk and meat, the commercial beef cattle industry remained confined primarily to the Black Belt of central and western Alabama. Anti–boll weevil measures and the cotton-based credit system assured the white fiber of its lofty position in Alabama agriculture for at least another generation.

Diversification became the key word for government agriculturists during World War I. Officials from Washington, D.C., down to those at the county seat told farmers it was their patriotic duty to plant more food crops and raise more livestock. In 1917 Alabama Cooperative Extension Service Director J. F. Duggar rang patriotic when he wrote, "He who does not diversify will help the Kaiser, and he who will grow something to feed the Allies will help whip the Kaiser." Extension agents were responsible for turning the rhetoric into reality. Not only did county agents encourage farmers to convert cotton acreage into food crops or pastures for livestock, but also they often played key roles in obtaining livestock for prospective breeders. In the summer of 1918 Marengo County agent Frank Curtis accompanied cattleman C. C. Clay on a trip to drought-stricken Texas, where they purchased a carload of beef cattle. After seeing the livestock, interested cattlemen and farmers around Demopolis sent Curtis and another agent back for twenty-five more carloads. Extension agents eventually arranged for the shipment of some 2,000 head of Hereford, Angus, Shorthorn, and Red Poll cattle into the Black Belt in 1918.[1]

Extension agents also arranged cattle purchases within the state. Working in cooperation with the State Council of Defense, the Alabama Cooperative Extension Service established a livestock exchange during the First World War. County agents made lists of prospective buyers and sellers and the number and kinds of stock they needed or offered and distributed the lists among cattlemen and other agents. Because the exchange appears to have favored purebred stock and few banks of the time offered loans for livestock purchases, the supply may have overwhelmed the demand. One list from October 1918 noted four buyers seeking a total of ten head along with seven sellers offering more than sixty head.[2]

The extension service, in addition to influencing wartime diversification, benefited from the emergencies caused by war and the boll weevil. By offering panic-stricken and patriotic farmers information on modern, scientific agricultural methods, the extension service legitimated itself to growing numbers of once skeptical farmers and established a permanent and prominent role in local agriculture. However, the cooperative extension service also reflected racial and economic prejudices. Black agents generally worked with fewer funds and supplies than white agents. Furthermore, white and black agents tended to establish themselves in the circles of the larger, more successful farmers, who possessed the resources and abilities necessary for diversification or utilization of modern techniques. This practice, in effect, helped doom the majority of agriculturists, that is, landless tenants and small farmers, to lives of increasingly uncertain prospects and ultimately to banishment from the land. As historian Pete Daniel posits, "Had its [the extension service's] information been dispensed evenly, sharecroppers and small farmers could have developed a strategy to stay on the land."[3]

For most extension agents and agricultural experts diversification into commercial cattle raising entailed the purchasing of purebred stock. Consequently, representatives from national livestock organizations played an important role in the growth of cattle raising. Before World War I the importation of purebred cattle into Alabama had affected the state's herds very little and in most areas not at all. A handful of Black Belt and Tennessee Valley breeders, along with a select group of dairymen, had introduced thoroughbred Jersey, Hereford, Angus, and Shorthorn cattle into their herds. For the most part, though, Alabama's cattle were the descendants of the old native cattle of Spanish and British blood. Small and lean, native cattle were well adapted to the seasonal rigors of the deep South but held little allure for agriculturists seeking meatier animals and thus more money at sale time. Recognizing the potential for expansion in the region, major breed associations sent agents into the South before World War I; these agents helped form state associations during the war.

In March 1917 field representative M. A. Judy helped a group of cattlemen and other agriculturists organize the Alabama Aberdeen-Angus Breeders' Association at Demopolis. The presence at the meeting of several Alabama Polytechnic Institute (now Auburn University) representatives, including C. A. Cary, underscores the cooperative relationship between government personnel, farmers, and private organization representatives. In November of the same year at the third annual state livestock show in Montgomery, cattlemen organized the Alabama Hereford Cattle Breeders Association. The officers and board of directors of the new organization represented a roll call of early, prominent cattlemen—N.J. Bell, J. E. Dunaway, R. J. Goode, Jr., W. G. Henderson, and Will Howard Smith. The association's leadership also reflected the dominance of Black Belt cattlemen; of the seven board members only L. H. White of Huntsville came from outside the region.[4]

Support for the cattle industry came from other sources as well. At the Alabama Live Stock Association's annual meeting in February 1917, N. J. Bell, a Lowndes County cattleman, delivered a lecture on beef cattle. Another participant added a speech encouraging cattle raising in the Tennessee Valley. Private individuals and businesses also provided impetus. In the spring of 1917 Marengo Farms of Demopolis conducted its first annual cattle sale, entertaining buyers from as far away as Tennessee. Manager Oscar E. Cobb of the Southern Cattle Company used newspapers to advertise his stock, offering 3,000 to 5,000 head of cattle for fattening. Furthermore, commission merchants at Louisville's Bourbon Stockyards asked cattle raisers to ship their stock by rail to the large Kentucky market.[5]

This last reference reveals a formidable barrier to the widespread growth of commercial cattle raising. Before World War I there was no terminal livestock market in Alabama, or in the Southeast for that matter. Cattlemen and farmers wanting to sell surplus cattle faced several options, all of which generally translated into less money than could be earned at a livestock auction. Smaller farmers often sold their stock to local butcher shops, country buyers, or traveling commission buyers at whatever price the buyer offered. Country buyers, usually local farmer-businessmen themselves, purchased cattle from area farmers and then shipped them by rail to markets in Louisville, St. Louis, and elsewhere. Commission buyers worked for packing houses in Birmingham, Atlanta, and other cities, buying stock from farmers across the region on commission to ship back to the packing house. Larger cattlemen frequently enjoyed the option of skipping the buyer and shipping livestock directly to Louisville or New Orleans for market prices.[6]

The steady expansion of cattle raising before and during World War I in the Black Belt and the potential for tremendous growth influenced a group of investors from Louisville and Montgomery to establish a terminal market in Alabama. In 1917 six stockholders made plans for the Union Stock Yards in Montgomery. Four of the six were Louisville livestock commission buyers—F. H. and G. W. Embry of Tatum-Embry Co.; W. D. Carrithers of Watkins, Carrithers; and William L. Kennett of Kennett-

Murray. The two Montgomery stockholders were William H. Teague and A. C. Davis. The stockholders elected Teague president and Davis secretary-treasurer and hired Harry E. Snow, a former traffic manager at the Bourbon Stock Yards in Louisville, as manager.[7]

The Union Stock Yards opened its gates for business on the morning of June 5, 1918, on a thirteen and a half acre plot in north Montgomery. Opening-day festivities included a speech by noted Demopolis cattleman C. C. Clay, in which he stressed the war effort and the growing demand for beef, and a dinner for prominent stockmen and buyers. Receipts of 650 cattle, 250 hogs, and a few goats and sheep brought $75,000 during the first day of business, with buyers ranging from Swift and Company of Andalusia to the Cincinnati Abatoir Company.[8]

In the years following World War I, the Union Stock Yards served various functions for Alabama cattle raisers. The yards offered central Alabama cattlemen a convenient market with competitive prices and numerous buyers in one location, spurred the establishment of the Montgomery Abatoir in the early 1920s, and provided an incentive for many farmers to diversify their operations. In addition, through the work of manager Harry E. Snow the stock yards boosted cattle raising around the state. Snow made speeches to farmers all over Alabama influencing them to replace cotton with livestock and pastures. He also assisted the extension service's efforts to purchase and distribute purebred bulls. In the spring of 1929 alone, Snow helped ship into Alabama 156 registered beef bulls.[9]

The World War I era accomplished an equally if not more dramatic increase in commercial dairying. Between 1910 and 1920 milk marketed by Alabama dairymen increased from 3,397,425 gallons to 6,408,962 gallons. Sales of cream grew by 430 percent to 150,474 gallons, and butterfat production for market in 1920 was twenty-five times greater than that in 1910. Unlike beef farmers, dairymen were not overrepresented in the Black Belt. The early dairy industry flourished near urban markets. Large dairies often delivered their own milk to city and town customers. Urban areas also contained ice cream plants and creameries. Consequently, by 1920 Birmingham's Jefferson County dairymen accounted for more than one-third of the state's marketed milk. Mobile and Montgomery counties combined to supply almost one-fifth of the remainder of the state's milk.[10]

The dairy industry was not a new development in Alabama in 1920.

Marion County farmer J. K. Davis with his young Jersey bull, 1927. (Courtesy of ACES Photo Collection, Auburn University Archives)

Alabama agriculturists had long been concerned with improved milk production and had begun importing registered Jersey cattle soon after the Civil War. The Agricultural and Mechanical College of Alabama at Auburn had established its own dairy herd in 1888 with stock purchased from the Hood Farm in Lowell, Massachusetts. By the second decade of the twentieth century several Alabama farmers operated large dairies, including Perry County's J. Freeman Suttle and Birmingham's Robert Jemison. One Black Belt dairy recorded more than $4,400 in cream sales for 1912 along with sales of purebred Jersey cattle to forty-one different buyers from three states and even Nicaragua.[11]

The war effort boosted the production of dairy products, just as it did the raising of beef cattle. Again, the land grant college and extension service played key roles. The Department of Animal Husbandry at Alabama Polytechnic Institute had established the first creamery in Alabama three years before the United States entered the war. The wartime demand for dairy foods resulted in the establishment of several more creameries. By 1921 Alabama contained seventeen such manufactories located throughout the state, with two each in Birmingham, Montgomery, and Selma. In December 1917, the Marengo County extension agent helped establish the state's first cattle testing association, the Canebrake Cow Testing Association. Composed of prominent western Alabama dairymen, the Canebrake association worked in cooperation with the Selma Creamery and Ice Company. Two years later API dairy specialist William Hunt Eaton organized the Macon-Lee Jersey Bull Association, which immediately purchased four young bulls from a New York farmer. In Tuscaloosa County the county agent, a local banker, and API dairy specialist William Hardie cooperated with dairymen in the purchase of a carload of Jersey cattle. Hardie also assisted the Selma Creamery and Ice Company with the importation of two carloads of Jersey heifers and bulls from Ohio. At Alexander City, banker and creamery owner Harry Herzfield bought a carload of Holstein cattle from Wisconsin and sold them to area farmers.[12]

Representatives from API continued to encourage dairy expansion after World War I. In the early 1920s dairymen and Auburn agriculturists organized the Alabama Jersey Cattle Club in Auburn. Early active members included some of the state's most influential dairymen: J. J. Paine of Grand Bay, J. C. Beane of Huntsville, M. W. Hall of Midway, W. J. Burton of Oxford, and W. J. Forrester of Dothan. In the late 1920s extension agents began to form 4-H Jersey Cattle Clubs for farm children, with the first such club formed in Chambers County in 1928. The expansion of dairy farms, especially in the Tennessee Valley and hills of northern Alabama, would eventually provide a foundation for the establishment of beef cattle raising after World War II.[13]

During the years surrounding the First World War, the dairy industry was often only indirectly linked to the growth in commercial cattle raising.

Another major social and economic development of the time, however, was in many instances directly related to the expansion of cattle raising. The Black Exodus from the Alabama countryside during the World War I era involved a complex assortment of economic forces that pushed landless African-Americans off farms and plantations and pulled them to cities such as Birmingham, Montgomery, Detroit, and Cleveland.

At a time when the boll weevil threatened the lifeblood of the southern economy, when growing legions of tenants and sharecroppers realized dwindling crops and returns from exhausted and eroded lands, and when ramshackle schools and almost nonexistent health care provisions threatened the very existence of future generations, World War I stimulated the machines of wartime production in the nation's industrial cities. Their supply of cheap, immigrant labor cut off by the war in Europe, industrialists looked southward to the thousands of discontented black laborers. Soon the steady trickle of blacks riding the rails to southern industrial cities such as Birmingham and northern giants such as Detroit grew into a flood. Close to half a million blacks left the South during World War I, in addition to the thousands who migrated to southern cities from the depressed countryside. Over 800,000 more moved northward in the 1920s. The black population of Alabama decreased by more than 150,000 between 1910 and 1930.[14]

The reasons for these migrations are probably too numerous to count. Some migrants had family members or friends who had gone north before the war. Others set out on their own in search of better economic opportunities and an improved racial climate. In 1923 black extension workers conducted questionnaires concerning outmigration in twelve Alabama counties. Respondents offered more than a dozen reasons for leaving the farm, most of which could be grouped into social and economic categories. A few of the most often cited reasons included a desire for better schools, the lack of court justice in Alabama, unfair crop settlements and high rents, poor living conditions, lack of freedom and control over their own operations, lynching and mob violence, and a desire to join friends and family.[15]

The extension agents surveying these fifty-eight communities found 362 families preparing to leave the state and join the 947 who had already fled Alabama. More than 900 families had relatives living in the north in 1923, and only sixty-eight families had returned to their communities after living in a northern city. Those remaining in Alabama claimed benefits because of the loss of labor for white landlords and employers, including better wages and working conditions, more appreciation from their employers, more respect from landlords and more favorable rents, and even better schools and improved race relations.[16]

White reactions to the Black Exodus were initially mixed. Many planters, especially those who continued to rely exclusively on cotton, were alarmed at the loss of so much labor. It appears that the cooperative extension service, beset by such fears, encouraged black agents' attempts to quell the growing tide of outmigration. The 1917 annual report of the director of black extension work implies such goals. Writing for his white supervisors, Thomas Campbell remarked that "all of our agents have proven themselves loyal Americans by helping to allay the excitement and unrest among our people, occasioned by the so-called negro exodus and the gigantic war entered into by our Government."[17]

Eventually, however, progressive agriculturists attempted to view the exodus in a positive manner. According to historian James R. Grossman, many white agricultural reformers viewed black exodus optimistically, convinced that the results would be diversification and mechanization. The reformers combined the ideals of progressive agriculture with the theories of black racial inferiority. Since Reconstruction reluctant planters had been faced with the dilemma of having to rely on the labor of free blacks. As early as 1875 one Pike County planter had employed racist language in an article urging the replacement of cotton with livestock: "Instead of being at the mercy of a lazy, thievish, demoralized and impudent set of negroes, we would again establish our supremacy, and dictate our terms of employment or have their places filled with intelligent laborers of our own race." Though the agricultural reformers responding to the Black Exodus succeeded in softening the tone of their arguments, the sentiments remained rather consistent.[18]

One example of this brand of agricultural reformer was API's P. O. Davis. In a 1923 article in the *American Review of Reviews,* Davis presented his positive outlook on the Black Exodus. He offered the stereotypical view of blacks "being well suited to cotton farming" and blamed many of the South's agricultural ills on the ignorance of black tenant farmers. According to Davis, a combination of wise labor use and mechanization would rescue the cotton economy and bolster livestock production. Ironically, Davis blamed black farmers for failing to adjust to the boll weevil, though extension agents regularly consulted only those agriculturists with the financial means to diversify. As Grossman points out, demanding landlords and a restrictive credit system continued to trap black tenants and sharecroppers in the destructive pattern of cotton planting.[19]

Geographer Hazel Stickney observes that the labor shortage influenced the transition from cotton to cattle, but it is quite likely that this conversion often precipitated outmigration as well. Some Black Belt farmers began converting cotton acreage to pasture and forage crops even before the boll weevil's arrival. It is unlikely that these agriculturists faced a labor shortage; more likely their decisions made expendable hundreds of tenant and sharecropper families on plantations throughout the region. When World War I industry began to absorb surplus landless laborers as well as many tenants and farmers whose contributions were not yet surplus, large planters found it even more convenient and pressing to replace cotton fields with pastures, thus continuing a cycle of displacement. P. O. Davis cited the case of one Lee County farmer who had taken over a six-tenant cotton farm. Within three years the farmer had purchased two tractors, initiated a program of diversification, and reduced his tenant families by half. On larger cattle farms the displacement was more severe. Where twenty-five or fifty tenant and sharecropper families had tilled the soil, fields of cattle and forage crops required only a couple of full-time workers and a much smaller number of seasonal laborers.[20]

Regardless of the role of outmigration in the expansion of cattle raising during World War I, the sudden popularity and profitability of commercial beef or dairy farming was unmistakable. Fueling the interest of breed association representatives, weevil-threatened planters, and progressive agriculturists were the growing demand for beef and the ascending price of cattle. Between 1915 and 1918 the average price paid for Alabama cattle grew by more than two-thirds, from 4 cents per pound to 6.7 cents. Such prices lent the attraction of profitability to the progressive agriculturist's ideal of diversification (Table 3, appendix). Consequently, in 1920 Alabama's cattle population reached the one million mark for the first time and experienced a 12-percent increase over the preceding decade (Table 1, appendix). Ten

Baling hay with a stationary hay press on a Hale County farm, 1927. (Courtesy of ACES Photo Collection, Auburn University Archives)

Black Belt counties accounted for almost half the increase, revealing the stronghold of Alabama cattle raising.[21]

The replacement of cotton with cattle in Alabama began to gain national attention. In December 1918, the Chicago-based *Breeder's Gazette* featured an article on "The Future of Cattle Feeding in the South," in which author L. A. Niven concentrated on the Alabama Black Belt. Niven wrote confidently, "The south is destined to become as great a section for cattle feeding as the middle West." According to the article, Alabama's commercial beef production had doubled in the past decade, and the state had experienced a larger increase in beef cattle herds than any other in 1918.[22]

Along with those of most other agricultural products, cattle prices plummeted soon after the end of World War I. The war machine shut down, and the tremendous increase in cattle raising left Alabamians and other American farmers with a surplus of marketable cattle. Alabama cattle prices declined 43 percent between 1919 and 1921, eventually bottoming out at 3.2 cents per pound in 1924. Consequently, the average value per head of Alabama cattle declined from a wartime high of $38.50 to $16.10 in 1925. Alabama cattle raisers responded to the recession by reducing their herds, a practice that continued to glut the market for several years after the war. Alabama's cattle population fell below 800,000 in the late 1920s, a decline of more than one-quarter from World War I holdings.[23]

Despite the spread of commercial cattle raising and dairy farming during the decade before 1920, cotton remained king in Alabama, especially in regions outside the Black Belt. In 1924 Alabama farmers planted almost three million acres in cotton, an increase of 320,000 acres from the 1919 figure. By 1929 cotton acreage had grown to more than three and one-half million and neared pre–boll weevil estimates. Furthermore, in the decade after World War I production increased 83 percent to more than 1.3 million bales. Statistics from this period reflect the declining spiral of farmer prospects as a result of decades of monocultural cultivation and restrictive credit conditions. Increasing numbers of farm operators fell into the landless class of tenants and sharecroppers. By 1930 65 percent of Alabama's farmers tilled other people's lands. Despite extension service efforts for diversification, the vast majority of these found themselves trapped in a vicious cycle of overproduction of cotton, harsh overuse of the land, and dwindling returns because of declining market prices. Of those who had attempted to diversify their farming operations or convert their farms to cattle ranches, according to Stickney, only the most fortunate weathered the recession of the 1920s. All but the largest and wealthiest found themselves overextended and forced to revert to cotton cultivation or some other plan of agriculture. Most of those who survived were Black Belt planters "with large landholdings inherited from the plantation system and with sufficient capital to absorb their losses."[24]

Although the immediate post–World War I years witnessed a sharp decline in cattle numbers, the foundation for a permanent commercial cattle industry had been laid. The strength of this foundation was centered in the Black Belt. Experiment station personnel continued to conduct cooperative research with cattlemen in the region. Of significant interest were steer-raising experiments carried out on Kirkwood Plantation in Marengo County after 1925. Owned by New Yorker H. P. Shedd and managed by prominent cattleman U. C. Jenkins, Kirkwood Plantation contained 4,000 acres of pasture and hay in the heart of the Black Belt. These experiments with Hereford cattle, and others across the region, helped influence increasing numbers of Black Belt landowners to take up cattle raising. Extension agents also proved integral to the effort of building interest in cattle raising. In his 1924 annual report, extension livestock specialist J. C. Grimes lamented farmers' decreasing interest in beef, swine, and sheep raising; nevertheless, two years later the livestock specialist reported a number of county agents helping purchase beef bulls for their constituent farmers. Furthermore, he cited as proof of growing interest in cattle raising the importation of a carload of bulls into Montgomery for distribution in six western Alabama counties.[25]

Along with the influence of extension and experiment station work, rising cattle prices and slumping cotton prices spurred a steady comeback of cattle numbers in the late 1920s. On the eve of the stock market crash in 1929, cattle prices topped six cents per pound for the first time in a decade and cattle numbers reached 800,000 once again. Much of the increase took place in the Black Belt. In 1928 extension service livestock specialist R. S. Sugg noted an increase of beef bulls in "our Black Belt counties," indicating that "a great number of large landowners in this section intend to expand this phase of the livestock industry." Extension agents, according to Sugg, also began encouraging cattlemen to replace scrub, native cows with purebred beef cows. In 1929 the livestock specialist observed the influence of Harry E. Snow and the Union Stock Yards in the Black Belt's growing interest in cattle.[26]

The boll weevil and slumping postwar cotton prices left much Black Belt land idle. Consequently, industrious agriculturists in the 1920s found opportunities for purchasing abandoned lands and raising cattle. T. Whit Athey, Jr., and Dr. M. L. Crawford have recalled the similar experiences of their fathers. Athey's father, after leaving his job as a cattle tick inspector for the Bureau of Animal Industry in 1924, began buying yearling calves from farmers near his lands in southern Montgomery and northern Crenshaw counties. In 1927 he purchased a registered Polled Hereford bull from an Iowa breeder at the Alabama State Fair and later imported a carload of cattle from the same Iowa farmer. The elder Athey utilized both his sandy lands near Grady and Black Belt lands he had purchased fifteen miles north of his home. Athey drove the cattle to his Black Belt lands for spring and summer pasture; in the fall he drove them back southward for winter. Dr. M. L. Crawford's father, Roy J., took advantage of depressed land prices in the 1920s. Realizing the possible profits to be made in the cattle business, the farmer from the hills of northwestern Perry County began buying large tracts of Black Belt land near Hamburg in 1927, generally for less than fifteen dollars per acre. For the next dozen years Roy J. Crawford, like Athey, seasonally drove his cattle back and forth between his hilly lands and Black Belt lands.[27]

Merle Prunty, Jr., observes that after 1925 the Black Belt slowly ceased being a cotton region. This transformation, which began with the arrival of the boll weevil, awaited further governmental action, however. Before feeling the impact of New Deal acreage reductions, Black Belt agriculturists witnessed the establishment of the Black Belt Substation at Marion Junction, a station that would centralize and help disseminate scientific information on livestock raising and pasture and forage crop growing in the region. The Black Belt Substation provided a final block in the foundation of an infrastructure for the region's and the state's cattle industry.[28]

In 1927 the Alabama legislature approved a plan that would allow the experiment station to construct a system of branch substations in the five key geographic areas of the state. Researchers at all the stations were to work with popular crops as well as components of diversified agriculture. The Tennessee Valley Substation concentrated on corn and cotton in addition to poultry and truck crops. Horticulture and truck crops initially dominated efforts at the Sand Mountain Substation, although poultry and swine production eventually demanded more attention. Peanuts quickly replaced cotton and corn in Wiregrass Substation experiments, and early work at the Gulf Coast Substation centered on vegetable raising. The presence of commercial cattle raisers in the Black Belt and the decline of cotton planting influenced the original direction of the Black Belt Substation toward livestock production and the restoration of soil fertility.[29]

The experiment station established the Black Belt Substation in 1930 on a 1,100-acre tract of Perry County land at Marion Junction. Agricultural research at the substation began the next year under the direction of K. G. Baker. Keatley Graham Baker, a native Texan and graduate of Kansas State College, had served as an extension agent in Texas before coming to

K. G. Baker, original superintendent of the Black Belt Substation at Marion Junction, looks over calves before a sale in Selma, 1950. (Courtesy of ACES Photo Collection, Auburn University Archives)

Alabama in 1918. He worked as an Alabama extension agent and as an agricultural agent for the Cotton Belt Railroad before accepting the position as director of the new Black Belt Substation. Like C. A. Cary and Harry Snow, Baker became yet another non-Alabamian who exercised significant influence over the state's cattle industry through his position of leadership. Baker and substation personnel undertook large-scale pasture experimentation during the Depression, including work with white and crimson clover and Dallis grass. The substation also tested several forage crops: Johnson grass, oats, black medic, and Caley peas. Furthermore, Baker's work revealed the value of fertilizers for pasture and forage crops. Baker became a resident expert on and booster of beef cattle, and the substation offered an excellent venue for entertaining area cattlemen and conducting educational sessions on cattle raising.[30]

In her 1961 dissertation, geographer Hazel Stickney identifies 1930 as the starting point for a true "cattle economy" in the Black Belt. Supporting this claim, she identifies the two developments triggering this phenomenon. In addition to the founding of the Black Belt Substation, Stickney notes the importance of the establishment of the Selma Stockyard in 1929. The outgrowth of a small butchering and packing business, the Selma Stockyard was the first federally bonded livestock auction in Alabama. Founded by the Suttles family, who were also prominent cattle raisers, the stockyard offered an important service to local cattlemen. Not a terminal market such as the Union Stock Yards, an auction such as the Selma Stockyard provided a forum for competitive sales, where various buyers from packing houses and meat processors as well as other cattlemen openly bid for livestock on display. The auction yard then subtracted a commission fee from the seller's earnings.[31]

As Stickney also recognizes, once the foundation was in place, Depression era measures and government agents increased interest in cattle raising. Extension service policies and New Deal legislation initiated an agricultural revolution that World War II and postwar era technological and economic changes carried out. The cotton-reduction program of the Agricultural Adjustment Administration, the drought-cattle relief program, soil conservation efforts, and cooperation between the extension service and the Farm Bureau all combined to speed the transition from cotton to livestock that had started some two decades earlier. Once again, the process reflected blurred lines between public and private initiative and illustrated government agencies' traditional practice of serving the most prominent agriculturists.

In the years immediately following the stock market crash and the opening of the Black Belt Substation, plummeting cotton prices and joint extension service–Farm Bureau efforts reinvigorated central Alabama cattle raising, which had remained quite stagnant in the decade after World War I. In 1929 the Alabama Farm Bureau established a commission firm at the Union Stock Yards in Montgomery. Through the heat of the Depression the Farm Bureau Commission Firm served as a coordinating office for Alabama cattle raisers and extension service personnel. This coordination was largely handled by J. D. Moore, who embodied the complex relationship between public and private agriculturists and represented the agribusiness progressivism of Black Belt cattlemen. John Daniel Moore, a native of Bullock County and a 1912 graduate of API, began working for the Alabama extension service in 1927 after stints as a high school teacher, railroad agricultural agent, and extension agent in Tennessee. Though employed as an "extension marketing specialist," he worked out of the Farm Bureau Commission Firm offices in Montgomery, reflecting the close and often ambiguous relationship between the two agencies. In fact, Moore would later leave the extension service to accept a job with the Alabama Farm Bureau.[32]

Even before New Deal programs launched widespread diversification efforts, Moore and the extension service oversaw an intense program of beef cattle boosterism. In 1931 extension agents helped place more than 250 purebred Hereford and Angus bulls on Alabama farms, primarily in the Black Belt. In November of the same year a small group of Black Belt cattlemen, with the assistance of extension agents and the Farm Bureau, organized the Alabama Lime Belt Feeder Calf Association. Originally composed of twenty cattlemen from nine counties with combined holdings of 3,500 head, the association sought to produce calves acceptable to corn belt feeders and to organize sales and shipments of their calves. Extension agents in five counties also organized 4-H beef calf clubs for breeding and showing prize calves.[33]

Extension agents in seven Black Belt counties conducted extensive beef cattle campaigns in 1932. Each agent directed a meeting of leading cattlemen, businessmen, and bankers in his respective county for the purpose of encouraging the purchase of purebred bulls and taking orders for animals. The state extension office also mailed letters regarding the value of purebred bulls to prominent farmers and area newspapers. After the extension service completed its booster campaign, the Union Stock Yards held an auction of purebred beef cattle shipped in from Tennessee and Missouri. Over 300 farmers and stockmen attended the event, which resulted in the purchase of sixty-four bulls and over 100 cows. Extension agents around Alabama placed an additional 347 bulls with prospective cattlemen in 1932.[34]

In the meantime, from his Montgomery office Moore served as both marketing and purchasing agent for his firm, extension agents, and Alabama stock raisers. Moore frequently used his network of contacts to drum up business for the Farm Bureau Commission Firm and the Union Stock Yards. In 1932 he wrote several cattlemen asking them to sell their stock at Union. Moore even offered to send a truck from the stockyards to load the cattle of prominent Wilcox County planter and cattleman Stonewall McConnico. In August 1933, Moore and members of the Alabama Lime Belt Feeder Calf Association encouraged western Alabama cattlemen to end their practice of selling livestock to an Atlanta purchasing firm and to sell their cattle instead in Montgomery.[35]

More frequent were Moore's efforts toward buying and selling cattle for individual farmers. In August 1932, he advised two prospective Lowndes County cattlemen to purchase Shorthorns from a dealer near Winchester, Tennessee, an area with which he was familiar from his work as an extension agent. Earlier Moore had directed a St. Clair County farmer, who wished to purchase a Jersey bull, to contact Robert Jemison of Birmingham. Moore organized one purchase of four carloads of steers and heifers for two unnamed cattlemen in 1934. For this Washington County sale he even coordinated roundup efforts through two local cattle raisers. In early 1933, Moore arranged for a large shipment of registered Hereford bulls from a Missouri breeder. He then mailed letters to prominent cattlemen advertising the new arrivals. Through communication with southeastern railroad representatives, Moore advertised Alabama cattle to interested buyers in other states. From V. W. Lewis, livestock agent for the Atlantic Coast Line Railroad, Moore collected names and addresses of southeastern cattlemen, to whom he mailed advertisements for Black Belt cattle. Moore contacted buyers from Florida, North Carolina, Tennessee, and Kentucky and often mentioned specific Alabama cattlemen in his booster letters. In some cases Moore's activities surpassed mere boosterism. In 1932 he promised a visiting Kentucky cattleman free transportation and service "due to my position as extension marketing specialist."[36]

Moore exercised his greatest influence on the growth of cattle raising through his cooperation with county extension agents. As extension marketing specialist, he supplied information to county agents and helped coordinate sales and purchases. In the early 1930s Moore communicated with agents from Limestone to Mobile County. Most often agents contacted Moore for advice or information on marketing cattle for their constituents. Lauderdale County agent G. B. Phillips informed Moore of several farmers' intentions to sell their beef stock, to which Moore replied by supplying agent Phillips with current Montgomery cattle prices. He also presented the option of marketing the cattle in Louisville or St. Louis. Moore urged the Limestone County extension agent to arrange steer sales to local packing companies or to market cattle in Nashville. The marketing specialist also advertised prominent Alabama breeders to inquiring agents. When given specific requests for livestock, Moore often referred agents to Black Belt breeders such as R. Lambert and Sons and J. T. Stokely. Also, in his capacity as commission firm buyer, Moore requested supplies of cattle from county agents, especially in the piney woods of southwestern Alabama. He asked Baldwin County agent E. E. Hale to supply "not too common" steers for two and one-half cents per pound and sent a commission buyer to accompany Hale to the roundup. Likewise, Moore secured a promise for the Clarke County agent to supply 200 to 300 steers from a Jackson cattleman.[37]

Moore's correspondence further reflected the socioeconomic status of pre–World War II cattle raisers. Like the first generation of midwestern-style Alabama cattlemen in the World War I era, Depression-era cattlemen frequently represented the class of farmers with the largest landholdings and prominent businessmen. Moore's contacts included such large-scale Black Belt stockmen as R. J. Goode, U. C. Jenkins, and Stonewall McConnico. In some respects Moore sat at the center of a self-perpetuating circle of mostly large Black Belt cattle raisers. The marketing specialist generated thousands of dollars worth of business for several prominent cattlemen, many of whom participated in the Lime Belt Feeder Calf Association. Several of Moore's correspondents relied on something other than agriculture for their livelihoods. Among those writing Moore with cattle inquiries were Madison "Mat" Jones, owner of Greensboro Warehouse Company; S. D. Bayer, superintendent of Greene County schools; and the Haigler brothers, owners of a 3,000-acre plantation and Haigler Auto Company in Hayneville and described by Moore as "well-to-do" college graduates. Other cattle raisers and prospective cattlemen contacting Moore included numerous doctors, attorneys, manufacturing executives, and railroad agents.[38]

With Farm Bureau and extension service assistance, a group of these prominent Black Belt cattlemen formed a cooperative, the Livestock Producers Association. "Owned and controlled by cattle men in the black belt and adjoining counties," the LPA, headquartered at the Union Stock Yards, consisted of some of the area's largest and most influential stock raisers.

After a reorganization of the association in early 1933, the executive committee included Robert G. Arrington, Charles W. Rittenour, U. C. Jenkins, and Roy J. Crawford. The LPA marked an important step toward the cooperative marketing of cattle and offered a ready forum from which cattlemen could present their requests in the coming New Deal. Again, Moore played a critical role in strengthening the LPA by urging nonmember cattlemen to consign their cattle with the association.[39]

The year 1933 also signaled the beginning of a government-controlled program of diversification and encouragement for livestock raising. Cognizant of the myriad problems in the American countryside, President Franklin D. Roosevelt and his administration addressed farmers' problems that Herbert Hoover's voluntary programs had failed to solve. Among the New Deal's most important and lasting developments were programs for supply-reducing subsidies, non–staple crop loans, soil conservation assistance, and purchase and redistribution of drought-stricken, western cattle. Years of agricultural depression provided sufficient incentive for desperate farmers to comply with federal mandates, and the firmly entrenched cooperative extension service provided a ready army of disciples to deliver the federal gospel of diversification, conservation, and mechanization.

At the center of the New Deal farm program was the Agricultural Adjustment Administration, and the primary feature of interest to most Alabama farmers was the Cotton Section of the AAA. Because years of overproduction had forced cotton prices to unprofitably low levels, the AAA cotton program offered payments to farmers who agreed to take cotton land out of production, with the intention of thereby reducing the supply to more closely approximate demand. In 1933 this required the plowing up of thousands of acres of growing crops. In the first year of the program over 70 percent of Alabama's cotton growers participated, and by 1934 the state's total cotton acreage had dropped to 2,130,000, 40 percent less than the 1929 total. The AAA cotton program continually chipped away at the state's staple crop. By 1940 Alabama's total crop had fallen below two million acres.[40]

Coinciding with the cotton-reduction program's enticements to stop growing cotton were other New Deal agencies offering incentives for raising cattle and other livestock. The first of these incentive programs firmly to take root in Alabama was the production credit system, which offered invaluable resources for agricultural production other than cotton growing. Since Reconstruction an oppressive credit system that demanded the planting of cotton had stifled Alabama agriculture. Livestock raising demanded a system of extended credit. Whereas the furnishing merchant or landlord advanced the cotton farmer seed, fertilizer, and other necessities in the spring with the knowledge that the fall harvest would produce a crop, there was no convenient timetable for the prospective cattle raiser. By the time a farmer purchased cows, had them bred, and raised calves large enough to market, eighteen months to two years might have passed.

Though a few banks in the Black Belt had begun offering loans for cattle purchases shortly before World War I, by the 1930s most financial institutions continued to refuse farm loans that were not earmarked for cotton. In 1932 the Hoover administration attempted to address this problem with the formation of the Regional Agricultural Credit Corporation under the Reconstruction Finance Corporation. Despite a dearth of funds and a virtual lame duck status, Jesse Hearin, manager of the Montgomery branch of the government's Agricultural Credit Bank, alerted Alabama financiers to the need for livestock and non–staple crop loans in a joint meeting of the Southeastern Livestock Association and the Agricultural Committee of the Alabama Bankers' Association in late 1932.[41]

In May 1933, Congress took a major step toward revolutionizing farm credit services with the establishment of the Farm Credit Administration, which superseded the Federal Farm Loan Board. Anchoring the program was the production credit system. The Farm Credit Act established a network of twelve Production Credit Corporations that in turn were to organize local credit cooperatives. Through these local credit associations farmers were able to secure long-term crop and livestock loans from the Federal Land Bank and the Intermediate Credit banks. By early 1934, Alabama contained nine production credit associations covering the entire state. The largest and most influential credit cooperative in Alabama was the Montgomery Production Credit Association, organized in December 1933 by thirty-nine central Alabama farmers and planters. Serving twelve Black Belt and adjoining counties, the MPCA was composed of a number of noted cattlemen, including its director, N. J. Bell. In 1934 the MPCA received a federal allocation of $375,000, of which $229,000 was loaned to 374 farmers in the first year.[42]

The MPCA became an invaluable source of credit for the blossoming central Alabama cattle industry. According to a prominent cattleman and former extension agent, T. Whit Athey, Jr., the credit associations introduced the concept of long-term credit. A financially strapped cattleman who owned a substantial herd of livestock and who possessed sufficient feedstuffs to maintain his investment through the winter months was frequently exempted from his loan payment for another year. The effectiveness of this federal program encouraged private banks to emulate its practices. Over the years the farmer-investors gained more control over the association because of the original bill's plan for a gradual reduction in federal capital grants. By 1944 private farmers and businessmen held over half the association's stock. In the first decade of operation, the MPCA granted some 8,600 loans worth several million dollars at interest rates of 4.5 percent or less.[43]

A development in 1934 reinforced cattlemen's demands for credit and provided another incentive to the growing commercial cattle industry. Beef cattle were not designated commodities in the original Agricultural Adjustment Act of 1933. As a result of the effects of drought in the West, however, cattlemen accepted a federal control-adjustment plan in the spring of 1934. In May the AAA created the Drought Relief Service, which began purchasing starving cattle from distressed western ranchers the next month. Between June 4, 1934, and January 31, 1935, the Drought Relief Service bought over eight million head of cattle from twenty western states, Illinois, and Florida. Almost half this number came from Texas and the Dakotas. About one out of five animals was condemned by the AAA, and the rest were turned over to a number of agencies, including the Federal Surplus Relief Corporation, state relief organizations, and the Indian Service.[44]

The cattle purchased by the Drought Relief Service did little to improve southern cattle raising, according to historian C. Roger Lambert. Of the cattle shipped to the Southeast, most were placed on large farms to graze until slaughtering and others were shipped directly to processing plants. Though farmers offering to graze the animals received payments, federal officials avoided marketing the high-quality cattle to farmers and stock raisers. Many southern farmers hoped to purchase drought cattle or exchange their native cattle for the purebred ones coming from the western ranches. According to Lambert, Federal Surplus Relief Corporation and Rural Rehabilitation officials supported the idea of improving southern stock, but AAA officials, fearing such a plan's unpopularity in the drought-stricken states, refused to "subsidize competition for the cattlemen." As a result, "an opportunity to improve the quality of southern livestock was cast aside by agricultural officials." Despite the AAA's intentions, some Alabamians feared the influx of drought cattle threatened the southern market without offering the benefits of breed improvement. In a September 1934 letter to Secretary of Agriculture Henry A. Wallace, J. D. Moore expressed concern that Alabama cattlemen were being overwhelmed by relief cattle. Moore felt that the relief cattle supplied to central Alabama's primary buyers, Swift packing plants in Montgomery and Georgia, reduced demand for commercial cattle. In reply one of Wallace's assistants attempted to assure Moore that relief cattle were only used for relief purposes and were not sold commercially to Swift and other companies.[45]

Nevertheless, Alabama cattlemen did manage to acquire a number of western and midwestern cattle through private and public channels. In the summer of 1934, the Montgomery Livestock Producers Association began assisting farmers and county agents with the purchase of registered bulls from Nebraska and South Dakota. In a letter to the Barbour County agent, J. D. Moore reported that the bulls were selling for as low as sixty dollars per head and that the MLPA would arrange purchase and delivery of animals after a deposit of fifteen dollars per head. Throughout late 1934 Moore received numerous letters from marketers in Texas and the Midwest concerning the sale of cattle to southeastern farmers. Two years later when drought again threatened western cattlemen, Alabama extension livestock specialist R. S. Sugg received requests for thousands of cattle. By this time the AAA had established a system through which southeastern cattlemen could purchase drought-stricken animals. Under this system cattlemen filed requests with their county agents after arranging to pay for the livestock with cash or credit from a local bank or a production credit association. The extension service then sent the requests to the Surplus Commodity Corporation in Washington, D.C., which arranged for the shipment of cattle to the Alabama Relief Administration.[46]

The importation of western drought cattle also introduced the screw worm to Alabama cattle. A tissue-eating larva of a large, dark blue-green fly, the screw worm brings sickness and even death to its infected host. Imported western cattle spread the pest throughout the Southeast in 1934.

In Alabama the heaviest infestation appeared in the counties adjacent to Montgomery, the destination of the majority of drought cattle brought into the state. By the end of 1934, over 50,000 Alabama animals were infected in two-thirds of the state's counties. Congress appropriated funds to combat the pest in early 1935 and charged the Bureau of Entomology with supervising the eradication effort and administering funds. In response a State Screw Worm Control Committee, consisting of the commissioner of agriculture and leading Auburn agriculturists, was established to cooperate with the Bureau's program.[47]

While drought relief offered an immediate boost to Alabama cattle raising, the New Deal's soil conservation programs provided perhaps the most integral long-term and statewide impetus for agricultural transformation. Decades of monocultural agriculture and farming methods poorly suited to local conditions had resulted in acres upon acres of hideously eroded hillsides and depleted soils. Though extension agents and agricultural experts had struggled for two decades to encourage terracing, wise farming methods, and planting of nutrient-restoring crops, they achieved little headway until the 1930s. Both the Soil Conservation Service and the Tennessee Valley Authority provided Alabama farmers with financial and educational assistance in land reclamation and conservation projects. These programs usually entailed agricultural transformation and often resulted in the establishment of more cattle farms.

Most northern Alabama counties were included in the territory of the Tennessee Valley Authority. Created in May 1933, the TVA cooperated with the Alabama extension service and experiment station to restore fertility to depleted hill and valley soils and, thereby, to improve economic conditions in the depressed area. Combined efforts of the TVA and federal agencies in Alabama led to the establishment of demonstration farms in fifteen northern counties, on which federal officials and local farmers tested fertilizing, terracing, and other conservation methods. Terracing, especially, became of great importance in hilly and mountainous sections where years of intensive row cropping had left gullies and ditches on hillside plots. The extension service began encouraging the use of terraces even before the 1930s. By March 1931 extension service director P. O. Davis counted 569 Alabama farmers in thirty-three counties who were licensed to terrace fields. The hilly counties of Cullman and Clay contained more of these farmers than any other counties. TVA and Soil Conservation Service efforts further improved and popularized terracing, utilizing laborers from Civilian Conservation Corps camps to construct terraces in the middle and late 1930s.[48]

Terracing, however, generally implied a continued reliance on row crop agriculture. Most small farmers who took advantage of government terracing programs lacked the funds or knowledge needed to break away from cotton raising and thus continued to deplete the soil, though with considerably less erosion. The TVA attempted to address this problem through distributing fertilizers or encouraging the sowing of legumes and permanent pasture grasses. Building on research conducted by experiment stations and cooperating with the extension service, TVA officials worked to provide conservation-oriented alternatives to cotton production. In the 1930s the extension service distributed more than 5,000 tons of phosphate fertilizers for use on pasture land. As a result of the TVA's influence, by the late 1930s many northern Alabama farmers, especially in the river valleys, were harvesting large acreages of crimson clover, vetch, and other kinds of hay.[49]

The Soil Conservation Service eventually had an even greater influence on conservation efforts across the state. Established as the Soil Erosion Service under the U.S. Department of the Interior in August 1933, the agency was transferred to the Department of Agriculture in the spring of 1935 and renamed the Soil Conservation Service. The SES utilized Civil Works Administration and Emergency Conservation Works labor and, by 1935, established forty erosion-control projects around the country. Alabama's lone project, directed by R. Y. Bailey, was the Buck and Sandy Creeks Project near Dadeville in Tallapoosa County. By 1940 the SCS, using Civilian Conservation Corps labor, had established four demonstration projects in Alabama. In addition to the Dadeville experiment, state conservationist Olin Medlock directed projects near Greenville, Anniston, and Marion. The SCS and extension agents encouraged local farmers to participate in soil conservation programs, which included terracing, fertilizing, and planting perennial grasses. In April 1941 Alabama became the first state to place all its farmland within soil conservation districts.[50]

As historian Theodore Saloutos acknowledges, after 1935 "soil conservation became fundamental to long-range farm policy and was a primary part of the adjustment program." In 1936 Congress passed the Soil Conservation and Domestic Allotment Act. Under this new system farmers could receive federal payments for replacing soil-depleting crops with soil-conserving and soil-restoring crops. Farmers were also eligible for reimbursements for fertilizers used on pasture and soil-building crops. The Soil Conservation and Domestic Allotment Act, by bringing all crops under one program and by establishing state offices, brought better organization to the diversification effort and reached increasing numbers of agriculturists.[51]

The soil conservation effort was responsible for one of the most infamous developments in recent southern agricultural history: the introduction of kudzu. Indicative of the extension service's and experiment station's tendency to trumpet new developments before acquiring a complete knowledge of effects and results, the kudzu drive eventually became an embarrassment to many agricultural experts, and the clinging vine has become a symbol of the South challenging the cotton that it often replaced. Imported from the Far East, the green vine appeared to be a perfect plant for the soil conservation effort. In addition to its function as an antierosion device, kudzu could be used as pasture or hay. The man responsible for introducing kudzu to Alabama and the Southeast was R. Y. "Dick" Bailey, who became director of the Dadeville Soil Conservation Demonstration Project in 1934. Within a year Bailey encouraged the sowing of 2,000 acres of the vine around Dadeville. By the end of 1941 Alabama farmers had planted over 100,000 acres of kudzu, and

Perhaps the most visible legacy of progressive agriculture in the South, kudzu became a popular forage crop in the 1930s and 1940s. (Courtesy of ACES Photo Collection, Auburn University Archives)

much more followed throughout that decade. Soon, however, farmers learned to their dismay that when left ungrazed or uncut for substantial periods the clinging, perennial vine crossed fences, climbed trees, and in general smothered everything in its path. Combatting the persistent vine continues to perplex agriculturists. In an ironic *Progressive Farmer* feature naming Dick Bailey a 1941 "man of the year," the author smugly noted that many Alabama farmers originally believed Bailey's kudzu program "the largest-scale big fool thing that had ever been done in agriculture." Postwar developments injected those doubts with retrospective legitimacy as farmers and pundits decried the effects of the new southern crop.[52]

Kudzu aside, soil conservation efforts generally had positive effects on Alabama lands and Alabama farming. The cattle industry particularly benefited from federal payments spurring pasture and forage-crop growth. For many farmers who had been financially unable or unwilling to replace cotton acres with Dallis grass, white and crimson clover, lespedeza, and other grasses, soil conservation payments provided the incentive and credit to effect the transition. In August 1941 the *Birmingham News* reported on three Lowndes County farmers affected by federal programs. With the assistance of the county extension agent and the Central Alabama Soil Conservation District, Roland Young, Fred Holladay, and Ed Mealing all initiated commercial cattle operations using old cotton lands. Mealing divided his farm into cotton lands and pastures and forage crops for beef cattle; Holladay completely transformed his operation into a beef cattle farm. Young carried out the most intricate transformation. On a farm that had once grown 100 acres of cotton and 200 acres of corn, Young turned to raising beef and dairy cattle on 215 acres of pasture grasses such as lespedeza, kudzu, and crimson clover and harvested an additional fifty acres of oats and barley. Such developments were not restricted to the Black Belt. In subsequent months the same newspaper carried similar stories of farmers from Morgan County in the Tennessee Valley to Choctaw County in southwestern Alabama.[53]

The key factor behind the success of New Deal farm legislation was the federal presence already established by the experiment station and the extension service. Both agencies were firmly entrenched in the farming activities of most counties by 1933 and, therefore, provided able and respected messengers for federal programs. K. G. Baker's Black Belt Substation supplied valuable information on pasture grasses and beef raising to extension agents and directly to farmers. Baker periodically hosted "field days" at the substation for interested cattlemen and extension agents. His May 1934 field day, for instance, featured lectures on pasture grasses and steer grazing tests. The extension service became more and more receptive and helpful to would-be cattlemen in the 1930s. Extension service director P. O. Davis urged bankers to give first priority to livestock loans. In 1938 Davis reorganized the livestock program to enhance its effectiveness and placed R. S. Sugg in charge of swine, sheep, and beef cattle work.[54]

The Bankhead-Jones Act of 1935 provided funds for additional extension agents and experiment station researchers. For the first time states were able to appoint assistant county agents to oversee areas of agricultural development formerly given less attention. In Alabama, and specifically in Black Belt and adjoining counties, many assistant agents concentrated on the growing cattle industry. William H. Johnson, Jr., accepted a position as an assistant county agent in 1935; his duties were almost totally devoted to beef cattle promotion and he spent most of his time working with 4-H beef calf clubs. Upon graduating from API in 1939, T. Whit Athey, Jr., began a three-year stint as Crenshaw County assistant extension agent. Like Johnson, Athey devoted his efforts toward establishing 4-H calf clubs and instructing members in beef cattle raising and grooming. Because the bureaucratic requirements of the cotton program demanded first priority among most county agents, assistants such as Johnson and Athey provided a valuable service for cattle raisers and diversification efforts.[55]

The extension service also received a boost from the American Farm Bureau Federation. Founded in Chicago in 1920, the Farm Bureau had its beginnings in local farm organizations assembled for the purpose of supporting county agents. The organization maintained its close relationship with the extension service as it spread into the Southeast and other regions of the country. In 1931 Alabamian Edward A. O'Neal was elected national president of the Farm Bureau. O'Neal, described by historian Christiana McFayden Campbell as a "Southern aristocrat of the old school," inherited his family's Lauderdale County plantation after graduating from Washington and Lee College and attending both API and the University of Illinois. As Farm Bureau president, O'Neal stressed membership efforts in the Southeast and particularly in Alabama. Consequently, by the mid-1930s almost half the region's Farm Bureau members resided in O'Neal's home state. In 1936 the Farm Bureau launched a frenzied membership drive in the Southeast in which AAA production control committees were invited to organize local Farm Bureaus. Alabama proved especially adept at the mission. Under director L. N. Duncan the extension service sponsored the organization of Farm Bureaus, and several Alabama extension workers were granted leaves of absence to assist the Farm Bureau in other southern states. The Alabama Farm Bureau's growing membership—from 2,590 in 1933 to 41,014 in 1943—introduced increasing numbers of farmers to federal conservation and control projects. At the same time the interwoven and ambiguous relationship between the extension service and the Farm Bureau created considerable confusion for farmers, many of whom believed membership in the Farm Bureau was required by the AAA.[56]

The administration of federal programs through established channels gave a distinct direction to the growth of cattle raising during the Great Depression. Extension agents had gained respectability and influence during the World War I era by aligning themselves with the county's major agriculturists and influential businessmen. That alliance carried over into the 1930s; consequently, improvements in land conservation and livestock raising among small, inefficient farmers remained the exception. However, the extension service's failure to serve the lower orders is not surprising in light of their businesslike approach. As Gilbert C. Fite observes, land grant leaders viewed farming as simply another business—a business in which the small and weak would not and perhaps should not survive. To these proponents of agricultural progressivism "the reality of southern agriculture [was] productivity and business efficiency."[57]

The methods of business efficiency and productivity proved effective as Alabama's cattle industry grew steadily in the second half of the decade. An extension service annual report for 1935 reported profits among beef owners for the first time since 1930. That year also witnessed a record for cattle receipts at Union Stock Yards; the 176,786 cows and calves received easily surpassed the previous high set in 1927. Alabama stockmen also purchased more than 250 purebred bulls in 1935; the extension animal husbandman, with financing from the Union Stock Yards and commission agent Nolan Huddleston, selected and shipped 107 of these animals from

Oklahoma. In the following year approximately 4,500 head of Hereford cattle and several carloads of Shorthorns were shipped in and distributed among farmers in twenty different counties.[58]

Stockmen from around the state organized the Livestock Growers Association in 1936, which was able to secure passage of several new livestock laws. The state assembly enacted laws requiring licenses for livestock dealers and providing for the registering of brands. Furthermore, the Alabama State Board of Agriculture adopted provisions to prevent cattle rustling. The year 1936 also witnessed a rejuvenation in the old practice of showing cattle. In April Union Stock Yards hosted a fat cattle show featuring more than 500 head. The 1936 Birmingham State Fair featured its first 4-H beef calf show as well. The following year saw four large fat cattle shows displaying almost 800 head. The Tri-States Fat Stock Show in Dothan stimulated interest in cattle raising among southeastern Alabama farmers. In addition, livestock shows began to take center stage at many county fairs, and some counties initiated 4-H calf shows.[59]

Extension agents helped place over 700 purebred beef bulls on Alabama farms in 1938 and increased the number of 4-H beef club members to 705 in thirty-eight counties. That same year witnessed the reestablishment of the Alabama Hereford Breeders Association and the Alabama Angus Breeders' Association. Extension agents played a key role in the formation of these associations. The largest and most influential association, the Alabama Hereford Breeders Association, was restored at a Montgomery farm in October by agents and cattlemen from eighteen central and southeastern counties. Both associations sponsored breed sales in the spring of 1939. In 1940 six towns hosted 4-H livestock shows, including first annual programs in Union Springs, Monroeville, and Talladega. Furthermore, by the end of 1940, county agents conducted beef cattle demonstrations in forty-eight different counties.[60]

The opening of several livestock auctions provided another impetus to the growing cattle industry in the 1930s. Until late in the decade these auctions were mainly confined to the Black Belt. At the end of 1937, the extension service counted four auction markets—one in each of the Black Belt towns of Selma, Demopolis, Epes, and Linden. By 1940, according to Extension Animal Husbandry Specialist William H. "Mutt" Gregory, Alabama contained fourteen auction yards, or sale barns. In many cases local farmers financed the construction of yards, sheds, and pens, which they then leased to a private operator who would conduct weekly or biweekly sales. Local livestock auctions stimulated cattle raising by providing convenient places to market cattle as well as a handy gallery for purchasing stock.[61]

By the latter years of the Depression thousands of Alabama farmers had begun raising herds of commercial beef cattle. These herds ranged in size from fewer than ten to hundreds of cattle. Many cattlemen labored in the early stages of agricultural transition, while others, especially in the Black Belt, owned herds that had been established decades earlier. Most maintained rather diversified farms. Bullock County farmer M. W. Hall, Jr., expanded a dairy-cotton farm with the purchase of ten registered Polled Hereford cows in 1935. The Hall farm, which was one of the earliest and most successful dairy operations in Alabama, would remain diversified through World War II before gradually focusing on beef cattle. In 1938 the best Hereford herd in the state, according to extension husbandman R. S. Sugg, belonged to Matthews cattleman A. C. Hartley. Hartley owned 5,000 acres of Montgomery County pastureland and a herd of more than 100 registered Herefords. Macon County planter and stockman W. E. Huddleston reflected the activities of many Black Belt agriculturists. On his 5,600-acre plantation near Tuskegee, Huddleston maintained eighty-five tenant families and 1,000 acres of cotton on the non–Black Belt soils. On his prairie lands, however, he raised 600 head of cattle—about 100 head of Holsteins for dairy production and a commercial beef herd of 500 Hereford and crossbred cattle.[62]

The story of Edward Wadsworth presents a familiar and interesting entrance into the commercial cattle business. Sons of a cotton farmer, Edward and his two brothers decided to diversify their Autauga County farm when they assumed control of operations in 1932. Upon winning a pig-raising contest that year, Edward used his $150 prize to purchase thirty-two steers. Though Wadsworth barely broke even when he sold the steers in the spring of 1933, the brothers obtained a loan from the New Orleans Agricultural Credit Bank for the purpose of making their first purchase of Hereford brood cows from a Florida breeder. Using Johnson grass—which had been choking out cotton crops before 1932—for hay and utilizing suggestions from government and railroad agricultural agents, the Wadsworths realized a modest profit in 1934 and continued to expand their cattle operation in the late 1930s. By World War II Edward had become one of central Alabama's leading cattlemen, and Wadsworth Farms had expanded to produce an array of crops and livestock, including cotton, corn, small grains, hogs, and cattle.[63]

One of the Black Belt's largest diversified plantations, the Webb Brothers Plantation in Perry County, demonstrated the influence of federal programs and national forces. Descendants of antebellum and postwar planters, J. C. and C. A. Webb, Jr., began transforming their 22,000-acre plantation when they assumed control of operations in 1932. The Webb brothers soon turned the plantation, once the exclusive domain of cotton and corn, into a diversified farm featuring beef cattle and other livestock, pastures, forage crops, cotton, corn, and timberland. Starting with fifty head of common stock cattle in 1932, they built the herd to 500 by 1941, 100 of which were registered Herefords. Throughout the Depression years the Webbs marketed dozens of cattle in Louisville, Montgomery, and Selma and sold livestock to packing houses, independent buyers, and other cattlemen. Interestingly, in 1932 and 1933 C. A. Webb, Jr., sold twenty-three cattle to Judson College, part of the money from which was credited to his daughter's tuition bill. Webb animals dominated livestock shows on the eve of the Second World War, claiming grand champion bull honors at the 1939 and 1940 Alabama State Fairs in Birmingham. In 1940 the brothers sold 250 head of cattle, though they continued to keep their heifers in an effort to build the herd.[64]

Soil conservation programs affected decisions on the plantation. By 1941 the Webb brothers had decreased their cotton crop to 1,500 acres, and this was relegated to sandy, non–Black Belt soils. In addition to their 2,000 acres of corn and 250 acres of oats, the Webbs followed Soil Conservation Service recommendations by planting large acreages in crimson clover, vetch, cowpeas, and soybeans. Furthermore, the plantation also contained 335 acres of kudzu and twenty-five acres of sericea lespedeza, both first introduced by SCS representatives. The Webbs terraced 7,000 acres of pasture lands in the late 1930s and began applying nitrogen and phosphate fertilizers to hayfields of Johnson grass and oats. The API-educated brothers also experimented with new fertilization methods in conjunction with research programs at Auburn.[65]

The Webb brothers' other agricultural activities reflected a knowledge of university-related diversification programs. To utilize their 12,000 acres of timberland and idle tenant farmers in the winter months, the Webbs operated their own sawmill and cotton gin. They also maintained a private lake stocked with bass and bream from government hatcheries, small grain patches for feeding the quail released on their lands, and a grove of 125 pecan trees. Elsewhere on the plantation were fields of wheat and sweet potatoes, sheep, hogs, turkeys, chickens, mules, and saddle horses.[66]

Though the Black Belt continued to lead the state in cattle production, farmers in other regions also reacted to government programs by investing in pastures and livestock. The *Birmingham News* observed the beginnings of purebred beef cattle raising in counties such as Cherokee and Monroe. Commercial cattle raising had made considerable headway in the Tennessee Valley by the late 1930s. Huntsville farmer R. J. St. Clair raised a herd of fifty-five cows and Hereford bulls along with cotton and corn. The Madison County farmer used fertilizers in his pastures of white and hop clover, lespedeza, orchard grass, and bluegrass and combined seeds from fields of crimson clover and soybeans. Karl Edward Burgreen began raising cattle and crops on his Athens farm in 1935 and two years later initiated a soil conservation project on his lands.[67]

One future Alabama Cattlemen's Association president from the Tennessee Valley also entered the cattle business in the 1930s. Malcolm "Mack" Maples, like so many other pre–World War II cattle raisers, grew up in an old planter family with landholdings dating back to the early nineteenth century. The son of a Vanderbilt-trained, Limestone County doctor, Mack, while still a teenager, began overseeing the sharecroppers on his father's cotton lands along the Elk River in 1934. Realizing that the poor crop returns failed to cover taxes, Maples began to grow interested in his uncle's cattle. Joe Maples, also a Vanderbilt-trained doctor, owned a herd of Hereford cattle that he had begun raising a few years before. In 1936 and 1937 Mack bought ten purebred Angus heifers from a cattleman in Giles County, Tennessee, and soon afterward purchased a $250 Angus bull from another Tennessee breeder. Maples then worked with extension agents and Farm Bureau representatives to improve his land for livestock raising and, with his uncle, conducted crossbreeding experiments with Angus and Hereford cattle.[68]

On the eve of World War II Alabama was in the midst of an agricultural revolution. New Deal programs forced a move away from total reliance on cotton and toward diversification and soil conservation. Between 1929 and 1939, Alabama's cotton crop declined by 46 percent, from 3,566,498 acres to 1,930,560. In the decade after 1930, the state's cattle population increased by 31 percent to almost 900,000. Alabama also witnessed significant increases in the production of peanuts, forage crops, and poultry. A major goal of the AAA in the Southeast was the replacement of cotton lands with pastures. In 1939 only Alabama, Mississippi, and Florida had produced noticeable increases in pasture production, and Alabama was one of the leading southeastern states in the production of livestock.[69]

As we have seen, the Black Belt featured the bulk of expansion in the commercial cattle industry with the Tennessee Valley a second area of substantial increase in cattle raising. In 1934 J. D. Moore informed an agent of the Southern Railway that Black Belt cattlemen continued to purchase most of the stocker cattle coming through the Union Stock Yards. A 1941 Alabama experiment station bulletin noted that cattle were most numerous in the prairie lands of the Black Belt and least numerous in thinly populated and uncultivatable areas. Statistics bear out this observation. In 1940 fully one-third of the state's cattle population was found in the ten core Black Belt counties. Furthermore, in order of rank, these counties constituted ten of the twelve top cattle-producing counties in Alabama. Conversely, rugged hill counties such as Clay, Coosa, and Winston had scarcely experienced the cattle boom. Some comparisons reflect the regional discrepancies in cattle raising on the eve of World War II. Montgomery County led the state with a total of 49,251 head of cattle as compared with Cleburne County's 3,792, the lowest figure in 1940. Five other Black Belt counties counted more than 25,000 head while farmers in seven other hill counties owned fewer than 7,000 head. Cattle holdings in the Black Belt amounted to almost ten head per farm unit; many hill counties averaged fewer than half that number per farm.[70]

A number of factors, including cotton production control, extension programs, and rising cattle prices, fueled increased emphasis on cattle raising in the Black Belt as well as in other areas of Alabama in the final years of the Depression. Largely because of AAA production controls, the decade of the 1930s witnessed a 46-percent decrease in Alabama's cotton crop.

Particularly affected by this decline were the Black Belt and Piedmont regions, both of which experienced declines of more than one-half their 1929 crops. By 1940 extension agents in forty-eight counties had initiated beef cattle production demonstrations, and thirty-eight counties had 4-H beef clubs. The counties not participating in beef promotion were by and large the hill counties of northeastern and northwestern Alabama, whose farmers lacked the resources of capital and large landholdings required to enter the commercial cattle business on the scale preferred by extension service leaders. Alabama's cattle fortunes also received a boost from rising prices, which, after bottoming out at 2.4 cents per pound in 1933, steadily improved to 5.1 cents in 1940.[71]

As a result of these influences, the extension service's annual animal husbandry report for 1940 reported an increased interest in beef cattle raising, especially in the Tennessee Valley and the Wiregrass region. Several counties in these two regions witnessed cattle population increases in excess of the state's average of just under one-third. In the Tennessee Valley, Limestone County's cattle population increased by more than 40 percent between 1930 and 1940, and Colbert County's 59-percent increase ranked first in the state for the same period. Wiregrass counties Dale, Henry, and Coffee all surpassed the state average in the growth of cattle numbers as well.[72]

Despite the growing popularity of cattle raising in the Tennessee Valley and Wiregrass region, by World War II the "open grazing areas of Southwest Alabama" trailed only the Black Belt in cattle production. Though open-range cattle lacked the thoroughbred breeding and scientific care of Black Belt cattle, they often served as a more important source of income to their owners than did livestock in the plantation districts. Baldwin and Monroe counties each contained more than 20,000 head of cattle in 1940, and Mobile and Washington county herders owned an average of almost ten head, well above the state average. In the piney woods and canebrakes of this region vestiges of the Southeast's traditional herding culture continued into the mid–twentieth century. As a young man in Covington County, James Cravy participated in roundups and cattle drives with his uncle and father in the 1930s and early 1940s. Using whips and dogs the riders gathered their range cattle into one of their cowpens for separating and selling each fall. The Cravys, who according to James were unique in their reliance on old herding techniques by World War II, utilized the open-range lands of southern Covington County, especially the thousands of acres owned by the Jackson Land and Lumber Company.[73]

The Cravys represented one of the last surviving remnants of a centuries-old practice of cattle raising. In September 1939, the Alabama legislature passed a law closing the range throughout the entire state; the law offered local beats the opportunity to enact legislation maintaining the open range, a policy that would be overturned a dozen years later when the organized forces of modern agriculture effectively ended the state's open-range tradition. The Cravys and other farmers who used this anomalous method faced an uncertain future in which the only certainty lay in the inevitable demise of open-range herding and the triumph of midwestern-style cattle raising adapted to southern conditions.[74]

On the eve of World War II a new agricultural industry had taken shape in Alabama. Led by former cotton planters of the Black Belt, agriculturists from around the state had begun to adopt the system of cattle raising first developed in the Midwest and characterized by the use of purebred, British beef breeds, barbed wire enclosures of pasture grasses and forage crops, scientific herd management, and year-round care. Because of land and capital requirements, this system generally excluded all but the largest landholders or particularly industrious farmers. Furthermore, these early cattlemen adapted this system to the southern climate and southern traditions.

Since the eighteenth century Alabama's cattlemen had depended on vast expanses of land for supporting cattle herds. Despite the steady demise of the open range in the first half of the twentieth century, the demand for land availability remained. Because large, profitable commercial cattle operations required hundreds and often thousands of acres of pasture and forage crops, the introduction of midwestern-style cattle raising to Alabama was generally limited to large landowners. The former cotton planters of the Black Belt and Tennessee Valley met this requirement. Furthermore, buying purebred livestock and sowing pasture grasses required outlays of hundreds and thousands of dollars, which again limited cattle raising among smaller agriculturists, especially before the institution of favorable credit services during the New Deal.

The midwestern-style cattle industry also took on characteristics of traditional southern agriculture. Climatic differences affected cattle raising. Mild winters caused a greater reliance on pastures and less emphasis on winter feed crops. The South's hot summers bred diseases to which the new British breeds were particularly susceptible. Most of the new cattlemen were also cotton planters or former cotton planters. Consequently, cattle raising frequently coexisted with cotton raising, and large cattlemen utilized black laborers to care for and transport stock. This practice hearkened back to the use of slaves as livestock handlers before the Civil War. Some cattlemen even maintained a few cotton tenant farmers to provide accessible labor for ranch purposes, regardless of the possible unprofitability of such limited cotton raising.

Until World War II cattle raising was largely restricted to former plantation districts and among large farmers in other parts of the state. The agricultural revolution had yet to transform the hinterland of Alabama. Again, war would set the wheels of change in motion. Less than two weeks before the assembly of Alabama closed the range in 1939, German troops overran Poland, initiating the six-year conflict that would accomplish unforeseen economic, social, and technological changes in the United States. Combined with the lasting effects of federal programs established by the New Deal, these changes would bring about agricultural revolution and farm depopulation such as the country had never known.

Chapter 5

Cattle in the Cotton Fields

The half century following the Great Depression witnessed an agricultural, economic, and demographic revolution unparalleled in the history of the American South. The forces of mechanization, urbanization, and government influence transformed a rural, agriculturally dependent section into a largely urban region no longer dependent on one or two staple crops and, for that matter, no longer chiefly dependent on farming. Cotton, which for almost a century and a half symbolized southern agriculture and society, lost its hallowed place as the king of the southern countryside. In its stead appeared the crops and livestock of a diversified farming economy, representing in some sense a realization of the dreams of generations of progressive agriculturists. Machines replaced animal and human muscle and sweat as new agricultural technology and mechanization revolutionized farm labor and activity. Just as the mule and horse became obsolete relics of a bygone era, so too did thousands of farm families. Consequently, in the three decades after 1940, in the words of historian Gilbert C. Fite, the South experienced "one of the greatest movements of people in history to occur within a single generation." And the trend has not changed. Over the past quarter century, though the rural population has rebounded slightly, the number of southerners and Americans living on farms has decreased each year.[1]

In the past half century the combination of government influence, technological development, and private organizational initiative has exacted tremendous changes on the farm and in farming communities. The constants over the years have been declining numbers of farms and increasing sizes of farms. Between 1940 and 1992, the number of farms in Alabama declined from 231,746 to 37,905, and harvested acreage fell from 7.1 million acres to 2.1 million. In one five-year span in the 1950s the state lost annually more than 12,000 farms and farm families. This trend has slowed only slightly. In the eighteen years after 1974, Alabama farm numbers experienced a reduction of one-third with over 1,000 farms being absorbed by neighboring farms or cities each year. In the process the average farm size in the state grew from fewer than 100 acres in 1940 to 223 acres in 1992.[2]

On the eve of World War II, Gilbert C. Fite observes, "the elements were present for a kind of agricultural takeoff." Mechanization, rising prices, existing federal programs, and wartime labor and military personnel demands began to change the face of agriculture, and, as Fite writes, "the most obvious change in southern agriculture was the shift to livestock." Over the three decades starting in the early 1940s, "millions of acres of eroded cotton land were planted to grass and trees." Like World War I, the Second World War provided a major impetus for the livestock industry and specifically for cattle raising. Unlike the First World War, however, World War II proved to be the starting point for a genuine agricultural revolution. Whereas in the years following World War I plummeting cattle prices and inexperience resulted in a significant decline in cattle raising, after 1945 favorable prices, government acreage reductions, and mechanization steadily directed farmers toward livestock raising and away from a reliance on cotton. The quarter century before 1940 had witnessed the laying of a solid foundation for the Alabama cattle industry. The combination of high prices, mechanization, federal programs, and massive outmigration sparked by World War II led farmers across the state to use that foundation in the construction of the modern commercial cattle industry.[3]

A number of public and private organizations and individuals contributed to the growth of cattle raising during the four years of the United States' involvement in World War II. Federal agencies such as the Agricultural Adjustment Administration and the Soil Conservation Service continued to accomplish the transition of cotton lands to pastures and forage crops. The cooperative extension service steadily spread the gospel of diversification into areas of the state outside the Black Belt and Tennessee Valley. Rising cattle prices and the efforts of rejuvenated breed associations and other cattlemen's groups exhibited the profitability of cattle raising to curious farmers.

As they had done a generation earlier during the First World War, extension agents urged production of foodstuffs for the war effort. By 1943 only four Alabama counties lacked extension programs in beef cattle raising. According to the director of animal husbandry these four counties were in an area of small, family-size farms unsuitable to beef cattle raising, or, in other words, areas devoid of the capital necessary to institute a program based on the extension service model. (As we shall see, this obvious preference for larger farms and financially successful farmers delayed the growth of commercial cattle raising in the small-farming hill counties.) In 1943 agents assisted over 7,000 farmers in some aspect of beef cattle raising and helped place over 1,800 purebred beef bulls on Alabama farms. Furthermore, by that date Alabama had 767 demonstration herds and 1,370 boys and girls in 4-H beef calf clubs. The extension service also cooperated with major breed associations to conduct sales of Shorthorns, Angus, and Herefords.[4]

One indication of the growing popularity of cattle raising during World War II was the expansion of the Black Belt Feeder Calf Sales. Beginning in 1939 with sales at Selma, Demopolis, and Epes, the Black Belt Feeder Cattle Producers Association, in cooperation with the extension service, the State Chamber of Commerce, and the farm products division of the Tennessee Coal, Iron, and Railroad Company, conducted an annual series of sales of feeder cattle, that is, calves weighing between 300 and 600 pounds. In 1942 the association added sales at Camden, Fort Deposit, and Union Springs; Montgomery joined the feeder calf circuit the following year. Held in the fall of the year, the Black Belt Feeder Calf Sales featured a series of shows in different towns on successive days. These sales attracted buyers from throughout Alabama and from as far away as North Carolina and Kentucky.[5]

The six sales in 1942 marketed a total of 4,497 calves at an average price of 11.81 cents per pound for a total of $266,695. Buyers included Zeigler, Eicher and Meninger, and other Alabama packers as well as visitors from five other states. The number of feeder calves marketed in 1943 decreased to 4,329, though the average price received increased slightly. Calves at Union Springs, Demopolis, Montgomery, and Epes brought more than twelve cents per pound. Demopolis, which by this time had become the center of the Alabama commercial cattle industry, accounted for more than one-quarter of the total sales. Both the president and vice-president of the Black Belt Feeder Cattle Producers Association, Selden Sheffield and Frank Spurlin, respectively, hailed from the Demopolis area, and E. P. Griffith's Demopolis Stockyards marketed on average about 1,000 beef cattle each week during the war years. In addition to the influence of the Black Belt Feeder Cattle Producers Association, the financial assistance provided by the Demopolis Production Credit Association spurred breed and pasture improvements in the western Black Belt. In Bullock County G. M. Edwards, president of the East Alabama Cattlemen's Association and the First National Bank of Union Springs, contributed to the growth of cattle raising by funding the importation of purebred bulls.[6]

Sales such as those in the Black Belt generated enthusiasm for cattle raising in addition to thousands of dollars. In 1943 Armour and Company representative Colonel Edward Wentworth, speaking to a crowd of buyers and sellers at the Selma sale, predicted big prospects for the cattle industry. After the war, he assured his listeners, Europe's demand for beef and live cattle to replenish their herds would prove a bonanza for American cattlemen. Responding to such promises from meat packers and urgings from extension agents, Alabama farmers purchased nearly 4,000 head of purebred beef cattle in 1944, and by the end of that year twelve counties boasted local cattlemen's associations with a total of more than 1,000 members.[7]

Newspaper accounts and extension service records reflected the growing interest in cattle raising around the state. The 1943 annual report of the extension animal husbandman noted the existence of commercial cattle-raising operations in all but five counties. Cattle raising steadily gained popularity in the Wiregrass region where farmers used peanut vines for hay. Farmers in southwestern Alabama planted corn after potato harvests in order to provide winter feed for cattle. The report also recognized growing cattle herds, both beef and dairy, in the Tennessee Valley. By the spring of 1944 Colbert County farmer A. L. McWilliams grazed 140 head of beef and dairy cattle on pastures and small grains planted with the assistance of the Tennessee Valley Authority, extension service, and Soil Conservation Service. Even some hill farmers took up livestock raising. The state's largest commercial cattle feeder in 1943, owning over 1,800 head, was Blount County's Houston Blackwood. Randolph County farmer Hugh Overton, who began seeding and fertilizing pastures of lespedeza and kudzu in 1940, grazed fifty-seven head of cattle and forty-four sheep in the spring of 1942.[8]

A key reason for the expansion of cattle raising during the war was the rising price of cattle (Table 3, appendix). Between 1940 and 1943 the market price of beef cattle almost doubled to an average of 9.9 cents per pound, the highest level since the Alabama Agricultural Statistics Service began compiling price statistics in 1909. After a slight dip in late 1943 and early 1944, beef cattle prices began a steady increase that would peak in 1951 at 24.6 cents per pound. High cattle prices, the availability of credit, and payments for cotton reduction and pasture growth enabled many farmers to make initial purchases of beef and dairy cattle or expand existing herds.[9]

The five-year period between 1940 and 1945 witnessed a tremendous increase in Alabama's cattle population. While cotton cultivation decreased by more than one-half million acres, Alabama's cattle population experienced a 30-percent increase. The number of mature beef cows almost doubled to 246,000, and the average value per head rose from $23.60 in 1940 to $41.40 in 1945. The highest percentage increases took place in southwestern and southeastern Alabama. Farmers in the four southwestern counties of Baldwin, Clarke, Mobile, and Washington increased their cattle herds by one-half or more. By January 1, 1945, cattle populations in Baldwin, Clarke, and Mobile counties had increased by more than 80 percent over their numbers on April 1, 1940. The other region of incredible expansion was the Wiregrass. Peanut farmers used livestock to finish off the leftovers of their harvests. Cattle herds in Barbour, Coffee, Covington, Geneva, Henry,

Houston, and Pike counties were among the fastest growing in the state during World War II.[10]

Farmers in other regions expanded cattle operations as well. In the Piedmont area Chambers and Elmore county farmers purchased dairy and beef cattle to graze their new fields of kudzu and lespedeza. A few hill counties experienced significant increases in cattle numbers, especially Walker and Blount, though the Appalachian region continued to lag behind other areas in cattle production. Cattle herds in the Tennessee Valley also grew during the war, with Lawrence and Lauderdale counties enjoying the greatest increases. The Black Belt, though outdistanced in percentage increase by southwestern Alabama and the Wiregrass region, continued to lead the state in cattle raising. In 1945 nine of the state's top ten cattle-producing counties were in the Black Belt. By that year cattle populations in four of the region's counties exceeded 40,000 head, and Montgomery County alone claimed over 70,000 head.[11]

As a result of the Black Belt's long-standing dominance of commercial cattle raising in Alabama, efforts to organize the state's cattle raisers during World War II originated in this region. In the first week of January 1944, the Demopolis and Alabama chambers of commerce, the Alabama extension service, and the farm products division of the Tennessee Coal, Iron, and Railroad Company sponsored the first Alabama Beef Cattle Show and Sale at Demopolis. The event, which featured prize Angus, Hereford, and Shorthorn cattle of breeders throughout central and western Alabama, appeared to TCI to be the perfect opportunity for encouraging the organization of the state's cattlemen. TCI's relationship with Black Belt planters extended back several decades. Since before World War I Black Belt landowners had supplied thousands of carloads of hay for TCI's working stock around Birmingham and northern Alabama. TCI had also long depended on the demand for agricultural products such as wire fencing, metal gates and troughs, and nails, as well as basic slag fertilizer. The continued growth of commercial cattle raising would only enhance profits from these products. Naturally, the company reasoned, statewide organization of cattle producers presented the best chance of ensuring the steady growth of cattle farming.[12]

Consequently, in late 1943 Luther Fuller, TCI's farm products division representative for Alabama, contacted U. C. Jenkins, who was president of the nearly defunct Alabama Livestock Growers Association, with a plan for organizing a state cattlemen's association. Jenkins and K. G. Baker, director of the Black Belt Substation and secretary of the ALGA, agreed to cooperate with Fuller. Fuller promptly drafted a letter on an ALGA letterhead, which Jenkins signed, warning cattlemen of a possible "roll-back" in beef prices because of wartime rationing by the federal government and inviting them to attend a meeting at the Demopolis Inn on the night of January 4, 1944. On December 29, 1943, the TCI representative mailed 332 copies of the letter to prominent cattlemen and extension service personnel.[13]

A group of nearly sixty cattlemen and county agents joined Fuller at the Demopolis Inn to organize the Alabama Cattlemen's Association. The actions of this group reflected the interrelationship between livestock growers, industry, and government and revealed the political approach this agribusiness group would take in the coming decades. The association from the beginning sought to gain a political foothold by electing as officers two prominent state senators. The group elected Black Belt planter and cattleman J. Bruce Henderson as the ACA's first president (Table 4, appendix). Reflecting a practice dating to Alabama's infancy, the members considered regional implications with their choice of Tennessee Valley agriculturist Robert J. Lowe as vice-president. The influence of Henderson, Lowe, and future leaders would ensure the fledgling group of an at times inordinate amount of political power.[14]

A closer look at the ACA's first senior officers reveals the exalted social and economic standing of the group's early leadership and of its membership in general. James Bruce Henderson, a native of Miller's Ferry in Wilcox County, was born in 1892 to a Scottish immigrant father and Canadian mother. By the turn of the century his father had amassed vast landholdings in the Black Belt. In 1916, after obtaining his B.S. and M.S. degrees in agriculture from the University of Illinois, Henderson returned to Miller's Ferry to join his father and brother in the management of a diverse operation that included cotton plantations, cattle, timberlands, and a mercantile. Elected to the state senate in 1938, Henderson also served as secretary-treasurer of the Planters Cotton and Cattle Credit Corporation and as president of the Plantation Saddle Horse Association of America.[15]

In 1944 Robert Joseph Lowe owned one of the Tennessee Valley's largest farming enterprises with holdings in excess of 2,000 acres. Born into an old, Episcopalian planter family in Madison County, Lowe attended the Virginia Military Institute. Upon his return to Alabama, he managed the family plantation and developed an interest in scientific cattle breeding. Elected to the state house of representatives in 1942 and later to the senate, Lowe became the ACA's second president in 1946 and a spokesperson and leader for cattle raising in his native region.[16]

As Dan W. Hollis observes, however, the three perhaps most influential men at the January 4 meeting were not even cattlemen. Luther Fuller, Tom C. Reid, and William H. Gregory, Jr., symbolized the new association's strong connections with industry, business, and government. As we have seen, TCI representative Fuller spearheaded efforts to assemble the cattlemen at Demopolis. Throughout the war years Fuller encouraged the growth of cattle raising in the Black Belt. In addition to sponsoring and directing a number of shows and sales, Fuller coordinated auctions at which midwestern buyers purchased livestock from Black Belt cattlemen.[17]

Fuller had for some time discussed the need for a state cattlemen's association with his friend William H. "Mutt" Gregory, Jr. Raised in Florence by a dentist father, Gregory received his B.S. degree in agriculture from Alabama Polytechnic Institute in 1928. After serving as an extension agent in Morgan and Cherokee counties, he joined the state staff in 1937 as extension agronomist. Three years later Gregory succeeded Dr. R. S. Sugg as director of animal husbandry for the extension service. Unlike Fuller, however, Gregory did not enjoy free reign to participate in the young cattlemen's group. Extension service director P. O. Davis, in part because of his friendship with Alabama Farm Bureau president Walter Randolph, saw little need for commodity organizations. Nevertheless, Gregory proved a tireless promoter of the commercial cattle industry. "Although Gregory and Luther Fuller never held elective office," writes Hollis, "these two men perhaps contributed more than any other two individuals to the early life and growth of the association."[18]

The third influential noncattleman was Tom C. Reid, the director of the Alabama State Chamber of Commerce's agricultural division. The cattlemen elected Reid secretary of the new organization, and as the ACA lacked a headquarters or office Reid consequently carried out his secretarial duties from his chamber office in Montgomery. This decision initiated an eighteen-year relationship between the ACA and the State Chamber of Commerce. By tying the fortunes of the fledgling organization to the established power of the chamber, Alabama's cattlemen ensured cooperation between agriculture, business, and government in the coming decades.[19]

World War II also provided the economic stimulus for the expansion of dairy farming in Alabama. Dairy farming in the state had developed slowly and steadily in the two decades before the Second World War. After the boom years of World War I, dairying had slumped temporarily before governmental and private efforts rejuvenated it. In the 1920s, with encouragement from extension agents, banks in some areas financed purchases of purebred dairy cattle and offered loans for planting feed crops. In 1929 the Tennessee Valley Substation at Belle Mina began research efforts on dairy production and pasturing dairy cattle. Numerous New Deal programs encouraged dairying. Among the key programs were the Tennessee Valley Authority and the Rural Electrification Administration. By supplying electricity to thousands of farm families, these programs allowed for the use of electric milk coolers and milking machines.[20]

Nevertheless, by 1940 dairy farming remained a relatively insignificant area of agriculture. Although the production of whole milk for market had tripled to 19.8 million gallons, the value of dairy products sold increased by less than 50 percent between 1920 and 1940. The primary barrier to dairy expansion on the eve of World War II was the lack of urban demand in Alabama. For those farmers who did produce dairy products for the market, possible destinations included cheese plants, creameries, ice cream manufactories, condensories, and the public whole milk market. Generally, the production of "Grade A" milk, or that which met the testing requirements set forth by the State Milk Control Board in 1935, was restricted to urban centers such as Birmingham, Mobile, and Montgomery and to the Black Belt where by 1941 four cooling stations furnished milk for markets in Birmingham and Mobile. Smaller dairymen unable to purchase the machinery needed for Grade A testing and farmers outside the Black Belt and urban areas marketed Grade B milk to manufacturing plants. By the beginning of World War II Alabama contained eight cheese plants, one condensory, and twelve creameries.[21]

Urban growth and military demands spurred dairy production during and after World War II. Between 1940 and 1945 the production of whole milk for market increased from 19.4 million to almost 36 million gallons. Over the five-year span the number of farmers selling whole milk increased by 119 percent to 17,384 while the value of marketed dairy products rose from $6.3 million to $15.4 million. The war years also produced a decided shift toward whole milk production. In 1945, 771 fewer farmers sold cream, and the amount of butter marketed had declined by one-fifth from 1940 estimates. Though most milk produced in Alabama in 1945 continued to be consumed on the farm, World War II accomplished an increased emphasis on market production. Fully one-quarter of total milk production in 1945 went to market, compared with only 15 percent of production in 1940.[22]

In the years following World War II dairy farmers increasingly turned to scientific feeding methods and purebred cattle to meet growing demands for milk. By 1950 over 2,000 Alabama dairymen produced Grade A milk, with the bulk of this production concentrated around urban areas and in the Black Belt and Tennessee Valley. Furthermore, almost 12,000 additional farmers supplied manufacturing plants with Grade B milk. Only Choctaw and Clarke counties in the southwestern piney woods produced no dairy products for market. Almost 2,000 farmers owned at least one purebred dairy cow in 1950. Of these over three-quarters were of the popular Jersey breed; Guernseys, Holsteins, and Brown Swiss accounted for most of the rest. The Tennessee Valley and Birmingham areas led the state in purebred dairying, though increasing numbers of dairy cattle began to be imported after 1950 in the Appalachian region and the Piedmont, where with the help of a General Education Board grant in 1943 the Piedmont Substation at Camp Hill had launched a program to encourage dairy farming.[23]

By the 1950s farmers in almost all parts of the state had access to urban milk markets or manufacturing plants. The expansion of dairying caused by World War II sparked the creation of several new establishments, which in turn created a demand for more dairy farmers. By 1950 the Black Belt contained a number of companies purchasing milk, including Southern Dairies in Montgomery, the Black Belt Creamery in Greensboro, Kraft Cheese and Blue Bonnet in Demopolis, and several smaller outlets. Farmers in northern Alabama found markets from Tupelo, Mississippi, to Cedartown, Georgia, and at the Boaz Creamery, Carnation in Decatur, Shoals

Cheese in Florence, and several other establishments. A closer look at two of the more successful dairy establishments in Alabama reveals the growth of the industry since the Great Depression.[24]

J. Lem Morrison and the Barber family rose to prominence from quite different backgrounds and experiences, though both ultimately controlled large Alabama dairy companies. In 1921 brothers George H. and Warren Barber moved to Birmingham from Wisconsin with a truckload of milk-processing equipment and founded Barber Brothers Dairy. Just one year earlier the Birmingham-Jefferson County Health Department had instituted a strict ordinance requiring pasteurization of all locally sold fluid milk. Because only a handful of Alabama cities had such milk ordinances in the early 1920s, few pasteurization plants existed; the Barbers promptly took advantage of this shortage. Throughout the decade the brothers purchased milk from about 300 family-owned dairy farms in the Birmingham area. George Barber, to ensure a steady supply of milk, even began arranging financing for farmers who imported dairy cattle from Wisconsin and Minnesota.[25]

Five years after selling their business to Foremost Dairies in 1929, George H. Barber and his son George W. purchased the Hansen Pure Milk Company and changed the name to Barber Dairies. By World War II Barber Dairies served supermarkets, institutions, and food services as well as the home delivery market. Again, during the war George W. imported seven carloads of cattle from Minnesota, which he distributed to his suppliers. Reflecting the postwar dairy boom, Barber's business expanded from nine to twenty-four routes by 1950. Barber Dairies experienced tremendous growth in the 1950s and 1960s. By 1970, when Barber opened a new twenty-seven-acre plant in Homewood, the company had purchased four smaller establishments and had expanded into neighboring states.[26]

Unlike the Barbers, John Lemly "Lem" Morrison was a native of Alabama and not raised with an intimate knowledge of the dairy industry. Raised in the sandy hills of northern Hale County, Lem Morrison left home in the mid-1920s and purchased a horse and a 100-acre Black Belt farm near Greensboro. In 1925 he bought 115 additional acres and acquired two Jersey cows, the cream from which he began selling to a local buyer. Five years later Libby, McNeil and Libby hired Morrison to manage their raw milk plant in Greensboro. In the late 1930s, while still employed at that plant, which had been purchased by Nestle, he began work as a field promoter for dairy farming. In addition to organizing milk routes in a six-county area, Morrison bought a 640-acre farm. When Nestle decided to close its Greensboro plant in 1942, he borrowed $4,000, bought the establishment, and reopened it.[27]

During the war Morrison marketed Grade A milk in Birmingham, Montgomery, and Mobile and enlarged his own dairy farm. By 1946, when his Black Belt Creamery opened in Greensboro, he owned 1,100 acres and 250 dairy cattle. In 1949 Morrison purchased Dixie Dairies of Mobile and the next year acquired the Dairy Fresh Company, whose name he applied to all his holdings thereafter. In the 1950s he cultivated dairy farming throughout western Alabama from Tuscaloosa to Mobile as an industry leader and as a representative and the first president of the American Dairy Association of Alabama. In the following decade Dairy Fresh bought smaller dairies in Greensboro, Dothan, Andalusia, and Enterprise, reflecting the expansion of the industry into the Wiregrass region.[28]

Dairy companies such as Barber and Dairy Fresh expanded according to the state's dairy farming development and provided incentives for dairy expansion through increased demand for milk. For many small farmers, particularly in the hills of northern Alabama, the acquisition of a small herd of dairy cattle marked the first step away from semi–subsistence farming and a dependence on the cash from a few acres of cotton. This early participation in the postwar dairy boom would also prove to be an important phase in the transformation from hillside row cropping to beef cattle raising that dominated the 1960s and 1970s.

More incredible and widespread than the expansion of the dairy industry was the growth of commercial cattle raising. In the 1950s cattle surpassed cotton as the state's leading agricultural commodity. Between 1944 and 1954, Alabama's cattle population increased by over half a million head and the total value of cattle on farms rose by $15 million. Over the same decade the average number of cattle per farm doubled to more than thirteen head. In 1954 Alabama farmers received $35 million from sales of cattle, almost double the $18 million received in 1944 and significantly more than sales from either poultry or hogs. Cotton acreage, on the other hand, declined from over 1.3 million in 1945 to less than 800,000 in 1959. Fertilizers, better breeds of cotton, acreage allotments, and mechanization allowed planters and farmers to raise as much or more of the white fiber as before on less land. The cattle industry often became the beneficiary of this decline as many farmers switched to livestock raising.[29]

As had historically been the case, the expansion of cattle raising in the decade and a half after World War II was not uniform throughout the state. Much of the growth took place in the Black Belt and the Tennessee Valley, with other significant increases in scattered counties from the Wiregrass to the Piedmont. In the Tennessee Valley, Limestone, Madison, and Morgan counties all experienced an increase in cattle numbers of more than 50 percent in the fifteen years after 1944. The Black Belt, with particularly significant increases in Autauga, Greene, and Lowndes counties, continued to lead the state in cattle production at the end of the 1950s. The decade also witnessed a decline in cattle populations in a dozen counties. The most significant occurred in southwestern Alabama where the state legislature's permanent closing of the range in 1951 ended two and a half centuries of open-range herding. The most conspicuous nonparticipants in the cattle boom, however, continued to be farmers in the Appalachian counties in the northern part of the state. In these counties the cattle population grew only slightly, or in some cases shrank, between 1944 and 1959.[30]

A comprehension of regional diversity has always been one of the necessities for an understanding of southern history. This is certainly true in the case of southern agriculture. Addressing postwar southern agriculture, historian Gilbert C. Fite writes, "It cannot be overemphasized ... that change occurred at vastly different rates in different areas and subregions of the South." Geographical differences have fostered a variety of regional agricultural practices within Alabama as well. After World War II the cooperative extension service identified eight different agricultural regions in Alabama, and each of these undoubtedly possessed local peculiarities and aberrations. We have already witnessed the genesis of modern cattle raising in the weevil-infested soils of the Black Belt and the demise of open-range herding in the piney woods and swamps of the southwest. While commercial cattle raising remained relatively uncommon among farmers in the Appalachian plateau, the Piedmont, and the upper coastal plain before 1960, it became a much more popular undertaking in the plantation districts of the Tennessee Valley and among the peanut farmers of the Wiregrass.[31]

Several factors contributed to the growth of cattle raising in selected regions of the state during the late 1940s and 1950s. The establishment of increasing numbers of livestock auctions both followed regional cattle increases and influenced other area farmers to take advantage of a local market by raising more livestock. Favorable cattle prices lured farmers from traditional row cropping into beef farming. The Alabama Cattlemen's Association expanded into all sections of the state, carrying along the booster spirit and continually building its considerable political sway. Postwar affluence and urbanization created burgeoning demands for beef and dairy products, and the Alabama packing industry expanded to help satisfy the demand. Perhaps the chief contributors to the postwar cattle boom, however, were governmental agencies such as the extension service and the agricultural experiment station.

Alabama veterinarians, extension agents, and agricultural scientists cooperated to bring common diseases of cattle under control after World War II. In 1947 the Alabama legislature approved a program to use state funds for the eradication of brucellosis (or Bang's disease), a sickness that frequently caused abortions and premature births among cattle and even death among mature cattle. State and federal authorities stepped up their efforts in the 1950s and eventually reduced cases of brucellosis to minuscule proportions by the late 1960s. The screw worm also continued to plague the state's cattle after the war. Extension personnel conducted campaigns against the pest until 1964 when federal authorities almost totally wiped out the screw worm by releasing millions of sterile flies over the Southeast.[32]

As they had before and during the war, extension agents frequently assumed a direct role in encouraging cattle raising. Extension animal husbandman Mutt Gregory, through newspaper articles and radio addresses, urged farmers to start beef cattle herds in anticipation of the return of crop production controls after World War II. In Colbert and other counties extension agents and representatives from API conducted instrumental meetings for farmers interested in commercial cattle raising. Furthermore, extension agents continued to assist farmers in the purchase of purebred livestock. In 1949 agents helped place 2,333 purebred beef bulls on Alabama farms and assisted in the purchase of more than 3,000 such animals four years later. Over that span the number of purebred herds in the state increased from 548 to 875, and by the latter year only one county contained no purebred herds. By 1953 more than one in seven Alabama farmers considered the sale of beef cattle to be one of their principal sources of income.[33]

County agents also operated through a number of more personal channels. Agents outside of the Appalachian counties designed production and marketing plans in cooperation with local chambers of commerce, banks, stockyards, and civic clubs as well as cattlemen. After 1948 director P. O. Davis allowed county agents to assist in the ACA's recruiting drive. In the postwar years agents greatly expanded their participation in 4-H activities. By 1953 Alabama contained almost 2,000 young 4-H beef club members. Cooperation with 4-H clubs also involved agents in the growing system of livestock shows. In addition to exhibits at county fairs, the 4-H and breed associations sponsored several annual shows throughout the state. For instance, in 1953 eight 4-H fat stock shows from Eufaula to Gadsden produced hundreds of prize calves bringing from twenty-one cents to forty-three cents per pound.[34]

The post–World War II cattle boom peaked in the early 1950s. Cattle prices increased steadily from 9.4 cents per pound in 1945 to 24.6 cents in 1951. After a slight decline in 1952, cattle prices plummeted by almost 40 percent in 1953. A number of factors accounted for this dive. Drought in the Midwest eliminated the demand for stocker cattle in many areas. The development of leather substitutes and artificial detergents began to take a toll on demands for hides and tallow. The beef industry faced increasing competition from poultry raisers, who had experienced even more phenomenal growth since 1945 than had cattle raisers. Most important, though, was the simple law of supply and demand. Over the course of only one year, January 1, 1952, to January 1, 1953, Alabama's cattle population expanded by an incredible 17 percent. Supply simply outdistanced demand in the mid-1950s. Not surprisingly, 1954, the year in which the state's cattle population reached its highest point until more than a decade later, saw the lowest average price since 1945.[35]

In contrast to the stagnation of the cattle industry after World War I, the decline of the mid-1950s proved to be short-lived. Much of the credit for the quick rebound belonged to the extension service, which continued to intensify beef promotion efforts. By 1958 the number of purebred bulls placed on Alabama farms had increased to almost 5,000 each year, and one

Above: Judges, spectators, and Polled Herefords at the Birmingham Fat Stock Show and Sale, 1949. *Below:* Christine Snellgrove with her reserve champion bull, Dale County, c. 1950. (Both photographs courtesy of ACES Photo Collection, Auburn University Archives)

year earlier the extension service had founded the Beef Herd Improvement Association, which promoted the breeding and raising of Hereford, Angus, and other thoroughbred beef cattle. The number of 4-H and extension service–sponsored cattle shows also increased to ten before 1960. Furthermore, as we shall see later, the late 1950s and early 1960s marked the first systematic extension service efforts to promote cattle raising in previously underserved areas, most notably the Appalachian counties. By 1959 Alabama's cattle population had dropped to 1.64 million and the average price had once again topped twenty cents per pound.[36]

Though the extension service familiarized farmers with cattle-raising techniques and programs, the agricultural experiment station generally supplied the information concerning scientific cattle farming. Among the key elements of cattle raising researched by experimenters were pasture and forage crop production and scientific breeding and feeding techniques. K. G. Baker initiated pasture experiments at the Black Belt Substation in the early 1930s, and by the late 1950s researchers at eight locations across Alabama studied a variety of pasture grasses, including Bermuda, Bahia, bluestem, crimson clover, rye, and sericea lespedeza. The 1940s witnessed the introduction of two vital grasses from outside of Alabama. In 1943 Glenn Burton of the Georgia Agricultural Experiment Station released coastal Bermuda for farm use. Coastal Bermuda, which thrives in hot, humid climates but suffers in areas that have temperatures commonly below fifty degrees Fahrenheit in the spring and fall, soon became a favorite pasture grass of farmers in southern and central Alabama. Northern Alabama cattlemen were likewise introduced to their grass of the future in 1946 when Edwin Jones brought back KY 31 fescue seeds from a trip to Pembroke, Kentucky. Researchers also demonstrated the necessity of fertilizing pastures.[37]

The substitution of pastures for row crops after World War II was one visual symbol of a larger agricultural revolution. According to Fite, "The most dramatic change in the rural landscape from Virginia to eastern Texas in the 1950s and 1960s was the millions of acres of grass and forage on which better bred cattle ranged." By 1960 Alabama had two and one-half million acres of white clover pasture alone, and farmers made use of another two million acres of unimproved pasture.[38]

Experiment station personnel had long conducted tests to determine efficient and valuable feeding practices. After World War II researchers increasingly studied the results of crossbreeding purebred cattle. In the late 1950s agricultural scientists at Auburn conducted a rigorous three-year experiment in crossbreeding different British breeds. By the early 1960s experimenters had also begun to crossbreed popular British breeds with heartier Brahman cattle in an attempt to develop meaty stock with disease- and heat-resistant characteristics.[39]

One barrier to the growth of cattle raising after World War I had been the absence of marketing facilities. Until the eve of the Depression Alabama contained no livestock auctions and only one stockyard. By the 1950s, on the other hand, livestock auctions abounded in the cattle-growing regions of the state. Growing cattle herds and the availability of better roads and trucks for transportation spurred the development of these auctions. After the Selma Stockyard was established in 1929, an Atlanta commission buyer built an auction at Epes in Sumter County. In subsequent years similar auctions were opened in towns across the Black Belt, including Demopolis, Uniontown, Livingston, Eutaw, Marion, and Linden. By 1940 Alabama counted fourteen auctions and stockyards, the majority of which were in the Black Belt. The post–World War II years witnessed the establishment of auctions in other regions, especially in the Tennessee Valley and the Wiregrass. By December 1951, there were seventy-five livestock auctions in Alabama, and over 90 percent of these held sales at least once per week.[40]

In the immediate post–World War II period Alabama farmers continued to have a choice among several sales options. In addition to attending livestock auctions, farmers could sell cattle to other farmers, country buyers, local livestock dealers, the terminal market in Montgomery, or local packers and butchers. By 1950 three out of four cattle and calves sold in the state were marketed at auctions. Slightly more than 10 percent of sales were made to country buyers and local dealers. Six percent of cattle for sale went from farm to farm, 5 percent were marketed at the Union Stock Yards in Montgomery, and 2 percent were bought by local packers and butchers. These numbers were not uniform throughout the state. While the vast majority of marketed cattle in the Black Belt and Tennessee Valley passed through auctions, few farmers in the Coastal Plains region utilized auctions and almost nine in ten head of Piedmont cattle were sold to country buyers and local dealers. Furthermore, many auctions in the hill counties and other areas with underdeveloped commercial cattle industries were quite small, often unlicensed, and frequently unable to secure competitive prices for cattle. In 1950 one-half of the state's auctions recorded sales of less than $500,000, and six generated less than $100,000; only twenty of the seventy-five auctions reported total sales in excess of one million dollars. A 1951 Alabama law placing restrictions and responsibilities on stockyards and subsequent acts dealing with such businesses brought about a sharp curtailment in purchasing activity by country buyers and local dealers and eventually threatened small, inefficient auctions. Nevertheless, the gradual spread of livestock auctions into areas throughout the state provided a significant stimulus to the conversion from row cropping to livestock raising.[41]

Another factor influencing the postwar expansion of commercial cattle raising was the Alabama Cattlemen's Association. The ACA also became intricately connected with the extension service. County agents frequently organized county-level meetings in the late 1940s and 1950s, a time when the ACA lacked the infrastructure and resources to accomplish such ground-level tasks. In some counties agents served as secretary of county cattlemen's associations, despite the initial objections of director P. O. Davis. In addition to the leadership of Mutt Gregory, the extension service provided both local and statewide leaders. T. Whit Athey, Jr., president of the ACA in 1955, was a former county agent, as was the ACA's first executive secretary, Edward Hamilton "Ham" Wilson. According to Wilson, "Had it not been for the Auburn University Extension Service . . . we would have [not] had any type of cattlemen's association that we have today in Alabama."[42]

Like the extension service, the ACA propounded the tenets of scientific and businesslike progressive agriculture. The organization had originated in the Black Belt, which had been home to such progressive cattle raisers since the World War I era, and its early leadership represented the region's chief proponents of midwestern-style cattle raising as adapted to the southern plantation culture. The group's business-minded goals were reflected in its eighteen-year affiliation with the State Chamber of Commerce. At the founding meeting in 1944, the ACA established a precedent by electing as its secretary the director of the state chamber's agricultural division. Furthermore, according to Ham Wilson, the state chamber recognized the benefit of strong ties with the ACA. The chamber, stated Wilson, "felt like the cattle industry was business people and many of the cattlemen owned big businesses." The ACA conducted its business from the Montgomery offices of the State Chamber of Commerce until 1964, and the chamber also paid the salary of the ACA's executive secretary for several years into the early 1960s.[43]

Early leaders of the ACA also reflected their organization's connection with progressive agriculture and business. Four of the group's first ten presidents held agricultural degrees from land grant universities. The majority of the presidents in the 1940s and 1950s (Table 4, appendix), in addition to their larger-than-average cattle operations, also held considerable business interests. O. J. Henley, the organization's fifth president, served as vice-president and general manager of Yellow Front Stores in western Alabama. The ACA's twelfth president, Mortimer H. Jordan IV, was a prominent Birmingham lawyer and a longtime vice-president of the Southern Natural Gas Company who owned a large Black Belt cattle ranch in Greene County. Arthur Tonsmeire, Jr., who served as ACA president in 1959, owned large farms in Mobile and Marengo counties but devoted the majority of his time to his presidencies of the First Federal Savings and Loan Association of Mobile and the American Federal Life Insurance Company. Other prominent businessmen who led the ACA included W. P. Breen of Greene County, M. C. Stallworth, Jr., of Washington County, and Carl B. Thomas of Madison County. Such leaders resembled the cattle industry's diversified, business-minded proponents of the World War I era. In the words of Ham Wilson, the presidents "were like the Blue Book of Alabama. They were . . . some of the most prominent Alabamians, not only [in] cattle but business."[44]

Acting on its progressive agricultural and businesslike beliefs, the ACA began to have a noticeable impact on agricultural legislation in the 1950s. In 1951 alone the ACA influenced the passage of four state laws related to the cattle industry. The organization was able to exert such political power through its influential leadership. Three of the group's first eight presidents were state senators, and the ACA was intricately connected with the powerful State Chamber of Commerce. One act provided for state aid for the promotion of agriculture through fairs and livestock shows. Another act outlined regulations and procedures for registering brands with the State

Department of Agriculture's Stockyards and Brands Division. This law was designed to prevent cattle rustling and dishonest livestock auction practices. Though the ACA sought a law requiring mandatory branding of all cattle, the act passed by the assembly made branding optional and levied an annual registration fee of five dollars.[45]

The other two acts reveal the standardizing elements of progressive, business-minded agriculture. The ACA fought for passage of a bill regulating the state's livestock markets. Designed to prevent unfair stockyard practices and to standardize livestock market procedure, the act required all markets to obtain a permit from the Commissioner of Agriculture and Industries. In addition the law demanded that all auctions be federally bonded and meet certain safety and sanitary standards. A similar law passed eight years later required all stockyards to maintain bonded weighers and gave the Commissioner of Agriculture more power to punish violators of existing laws.[46]

The fourth 1951 bill was signed into law on May 30, and though the condition that it legislated was in essence a fait accompli by 1951, it carried symbolic weight. This act permanently closed the range throughout the entire state. Two hundred fifty years of open-range herding came to an end. Though the law affected a minuscule percentage of rural Alabamians, it did create consternation among several cattlemen in southwestern Alabama. A few farmers and herders still ranged cattle in the vast timberlands and swamps of the region as their ancestors had done for generations. William Johnson, Jr., attended a meeting in Montgomery concerning the proposed stock law and remembered the objections of southern Alabamians in the "long and bitter meeting." Nevertheless, with ACA support and responses of apathy from other regions of the state, the bill passed and forced a major change in agricultural practices for some affected farmers. One result of the closing of the range was a decline in cattle populations in southwestern Alabama. Between 1945 and 1959, while more than four in five Alabama counties experienced cattle increases, Choctaw, Clarke, Escambia, Mobile, and Washington counties all suffered decreases in cattle numbers. Especially injured by the range closure were Choctaw and Clarke counties, which experienced decreases of 17 percent and 40 percent, respectively.[47]

Ironically, at the same time that the ACA fought to abolish the last vestige of traditional, open-range herding, the association and its leaders began to adopt the trappings of a bygone era in cattle raising. Unaware of or unconcerned with the southern roots of western ranching, Black Belt cattlemen began raising quarter horses and sponsoring rodeos after World War II. Among the descendants of planters and yeoman farmers, cowboy hats and horsemanship became symbols of successful cattle raisers, while the descendants of piney woods herders found themselves plowing behind their ponies and stretching barbed wire. Furthermore, by the late 1940s the association had adopted as its symbol the lean and wiry longhorn, despite the fact that this descendant of Spanish stock and its southeastern, piney woods relative represented everything abhorrent to modern, progressive cattlemen. Notwithstanding the embracing of these superficial elements by the ACA, the 1951 range law simply capped a decades-old effort to replace the piney woods cow and its traditional herder with the Hereford or Angus and its scientific breeder.[48]

In addition to the above accomplishments, the ACA used its political strength in support of a number of other causes. In 1948 the association was influential in securing the passage of a "coliseum bill." The act appropriated money for the construction of a state coliseum (now Garrett Coliseum) in Montgomery and the appointment of Tom C. Reid, former ACA secretary, as executive officer of the State Agricultural Center. In the 1950s the ACA succeeded in getting cattle exempted from ad valorem taxes and saw passage of a state bill declaring cattle rustling grand larceny.[49]

Much of the explanation behind the ACA's rapidly growing power lay in its unique organizational structure and the tireless campaigning of its first executive secretary. In 1948, in an effort to boost sagging rolls, Mutt Gregory and Luther Fuller spearheaded a movement to restructure the organization's membership. The new constitution, drafted by lawyer Mortimer Jordan IV, allowed for a system of county associations allied with the ACA. The five-dollar membership fee was to be divided evenly between the local group and the state organization. As Gregory, Fuller, and others had expected, the new, locally focused design realized immediate dividends. By May 1950 nine county associations had affiliated with the ACA; by the end of that year the ACA counted some 1,400 members, as compared with only 102 three years earlier. In 1954 the final six county affiliates joined the ACA and the statewide membership exceeded 4,200. By the end of the decade the ACA's 6,143 members made it the second-largest state cattlemen's organization in the United States. This rapid success Ham Wilson attributed solely to the county-level structure established in 1948.[50]

Others attribute much of the group's success to Wilson. Wilson, a Greenville native and 1943 agricultural science graduate of API, accepted the post of secretary of the State Chamber of Commerce's agricultural division in April 1952. By doing so he also became secretary of the ACA, a position he would occupy for thirty-three years. The following year Wilson's title was changed to executive secretary, and in 1962, with the separation of the ACA from the State Chamber of Commerce, he became the cattlemen's executive vice-president. In addition to his efforts toward increasing membership, Wilson directed a number of significant developments in the 1950s and 1960s. After attending the 1952 annual meeting of the American Cattlemen's Association, Wilson and Mack Maples helped organize a women's auxiliary, the Alabama Cow Belles Association. Wilson later played a key role in beef promotion through both the ACA and the Alabama Beef Council, and in 1958 he produced the first issue of the organization's magazine, *Alabama Cattleman*. In that same year he and five others formed the Southeastern Livestock Exposition, which began sponsoring the Southeastern World Championship Rodeo and Livestock Week.[51]

Wilson's most crucial struggle involved the issue of funding beef promotion. Through the efforts of Wilson and other ACA officials, the state assembly passed a Beef Promotion Bill in 1959. The act, known as the "beef checkoff" law, stipulated that a deduction of ten cents would be made on each head of cattle sold at an Alabama stockyard. The funds raised by the law were to be used only for purposes of promoting the raising of beef cattle and the consumption of beef. Not all cattlemen reacted positively to the beef checkoff program, however. Quite unexpectedly, state senator and first president of the ACA J. Bruce Henderson filed suit to have the law overturned. In February 1960 Montgomery Circuit Court Judge Walter B. Jones declared the law unconstitutional.[52]

According to Ham Wilson, Henderson represented a substantial number of "reactionary" Black Belt cattlemen who opposed government interference in the industry, despite the fact that the law allowed opponents to recoup their dimes if they wished. In addition, recalled Wilson, the Farm Bureau, fearful that a commodity organization would achieve a level of power and influence equal to their own, worked behind the scenes to stir opposition to the bill. Rather than conduct a lengthy court battle, the ACA encouraged the introduction of a constitutional amendment outlining the beef promotion campaign. After the state assembly passed this legislation, the amendment was approved by 65 percent of the voters in a 1962 referendum. Only three counties, including Henderson's Wilcox County, returned majorities against the amendment. In August 1962 a referendum among the state's cattlemen overwhelmingly approved the beef checkoff program. Beef checkoff funds supported beef promotion in Alabama for over two decades until a federal law established a national twenty-five-cent checkoff fee in 1985. Three years later the national checkoff was raised to one dollar per head.[53]

Despite the various political achievements of the ACA, as Alabama entered the 1960s the cattle industry had become prominent in only the Black Belt and other selected regions of the state. While the Black Belt and Tennessee Valley continued to lead the state in commercial cattle production, southeastern and southwestern Alabama had also witnessed considerable development of the industry by 1960. A look at the ACA's leadership shows that it mirrored the geographically expanding industry (Table 4, appendix). Through 1960 all fifteen ACA presidents had come from one of the four regions mentioned above. Though five of the first six had been Black Belt cattlemen, only three of the following nine were from this region, reflecting the steady diffusion of cattle raising and influence into other areas. Between 1952 and 1960, the list of ACA leaders contained three from the Black Belt, two Tennessee Valley farmers, two Wiregrass representatives, and two southwestern cattlemen.[54]

Nevertheless, the Black Belt remained easily the top cattle-producing region in the state and one of the best in the South. Statistics from the 1959 agricultural census reveal that farmers in ten Black Belt counties owned one-third of the state's cattle. The region contained the top seven cattle-producing counties in Alabama and ten of the first twelve. Furthermore, the ten Black Belt counties accounted for more than 35 percent of all cattle sales in 1959. Montgomery, Lowndes, and Dallas counties generated more than $3 million each in cattle sales.[55]

Cattle raising in the Black Belt, and in the plantation district of the Tennessee Valley and southeastern Alabama, involved a combination of scientific, midwestern methods and plantation agriculture traditions. The expansion of the livestock industry in these areas was intricately connected with the post–World War II agricultural revolution and black outmigration. By examining the development of a number of Black Belt cattle operations one becomes aware of the effects of government and state agencies, mechanization, science, and rural transformation on the cattle industry in this leading region.

In her dissertation on the movement from cotton to cattle raising, Hazel Stickney offers several valuable case studies of Black Belt farm developments. The case studies present a fairly broad spectrum of cattle raisers and a myriad of variables effecting the conversion from cotton to cattle. On the Hawkins farm in Sumter County, beef cattle had provided a rather insignificant portion of the income until 1937 when Mrs. Hawkins, responding to a self-described labor shortage, converted the farm almost exclusively to cattle raising. By 1960, under the ownership of her son William, the Hawkins farm contained some 175 brood cows and only five and one-half acres of cotton. According to Stickney, Hawkins maintained one black sharecropper to ensure the year-round availability of labor on the farm, a "widespread practice" in the region in 1960. In addition Hawkins owned two tractors and employed one additional wage worker.[56]

Elsewhere in Sumter County, the Larkin farm, which had been a cotton plantation with a dozen tenant families in 1930, was transformed into a cattle ranch after World War II. By 1960 Larkin owned a herd of 150 Angus brood cows. Reflecting a growing trend, Larkin received an outside income as a rural mail carrier and, like Hawkins, he relied on black labor to maintain his cattle business. The Fields farm presented a different background. The owner, who had purchased the farm of over 400 acres while a Civilian Conservation Corps field superintendent in the 1930s, transformed the abandoned land into a cattle farm during World War II. With the assistance of extension agents and agricultural research information, Fields planted several acres of Dallis grass, the seeds from which he sold each year. By 1960, unlike most Alabamians, Fields earned his cattle income not by raising calves but by "finishing steers." Each fall he bought about 200 "feeder" calves weighing between 300 and 600 pounds, which he fed over the winter and sold in the spring.[57]

In Dallas County, where cotton production remained more significant than in most other Black Belt counties, Stickney documents the development of several cattle farms. The Ellis farm offered a variation on Merle Prunty, Jr.'s neoplantation by 1960. In 1930, fifty black families had tilled and cultivated a total of 450 acres of cotton on a 600-acre farm. A few of the black tenants had also assisted the owner with his Grade B dairy operation. During the Depression Ellis began to purchase Hereford bulls, and Agricultural Adjustment Administration acreage reduction furthered the move toward cattle production. By 1960 Ellis owned or rented a total of 5,000 acres, of which only 182 were in cotton. The cattle herd contained 600 brood cows of various breeds and more than twenty bulls. Despite the thorough mechanization of the farm in 1960—implements included a cotton picker, five tractors, a field harvester, two combines, and a hay baler—Ellis continued to employ fifteen black families as sharecroppers and livestock workers.[58]

AAA allotments also took a toll on the Suttles plantation. In 1930 dozens of black sharecropper families had worked on more than 1,000 acres of cotton lands. By 1940 the AAA had cut Suttles's allotment in half, forcing a transformation to some other commodity. In 1960 the 4,300-acre farm contained only forty acres of cotton and four black cash-renter families. Suttles owned 400 head of Hereford brood cows and, depending on the time of year, up to 2,000 additional cattle.[59]

Autauga County farmer Edward Wadsworth and his two brothers began diversifying their father's cotton farm during the Depression by purchasing Hereford cattle. By World War II the Wadsworth brothers raised a variety of agricultural products, including cattle, cotton, grains, hogs, and pecans. After helping found the Alabama Cattlemen's Association in 1944, Edward became a proponent of improved, scientific cattle raising and governmental cooperation with farmers. By the late 1950s Wadsworth Brothers Farm included 4,000 acres of timberland, pastures, and crops and contained almost 400 brood cows. In 1958 the sale of calves alone amounted to almost $40,000 for the Wadsworths.[60]

Not all cattle raisers followed the path from cotton planter to cattleman, however. One of Stickney's Sumter County subjects was Carol Jones, a black farm owner. Jones had been a tenant farmer for twenty years before purchasing a large farm in 1941. By 1960 Jones raised 250 head of cattle and 600 sheep on his 1,248-acre farm. In addition he rented another 600 acres on which he grazed 75 brood cows and 150 steers. Interestingly, Jones maintained his allotment of twenty-eight acres of cotton for the employment of family members.[61]

As mentioned earlier, several agriculturists from outside the region settled in the Black Belt to raise cattle. In addition to Alabama farmers who relocated to the area, cattlemen from other states moved to central and western Alabama. During the cattle boom of World War I, the Wallaces from Kentucky and the Caleys from Ohio purchased farms in Dallas County. By 1960 the Wallaces had quit growing cotton and owned a large herd of registered Angus cattle. The Caleys, after dairy farming in the 1920s, utilized research from the Black Belt Substation to establish their own herd of Angus cattle. In 1951 John Armstrong, member of a successful Texas ranching family, purchased a 5,000-acre plot on the Alabama River in Autauga County. Armstrong soon converted half the tract to pasture and transplanted Santa Gertrudis cattle from the family ranch in Texas. By 1960 the Black Belt ranch contained 500 brood cows.[62]

By and large, however, the key figures in the Alabama cattle industry by the early 1960s resembled the original proponents of midwestern cattle raising in the state. The majority were well educated, involved in diverse business interests, and heirs to the plantation tradition. Selden Sheffield, a native of Wilcox County, operated a 3,300-acre cattle ranch in Hale County with a herd of more than 600 Hereford brood cows. After graduating from Howard College (now Samford University), he began his cattle business in 1938 and helped found the ACA six years later. By 1964, when he was elected ACA president, he was serving as a member of the Howard College Board of Trustees and as chair of the Federal Reserve Bank of Birmingham and was a past president of the Demopolis Rotary Club.[63]

Another Black Belt cattleman, G. W. "Billy" Robertson, owned the 5,000-acre Marbilou Ranch in Hale County. A graduate of API and West Point, Robertson was a director of the Canebrake Bank in Uniontown and president of the Black Belt Stock Yard Corporation. John M. Trotman, an API graduate and Montgomery Country Club member, ran a 7,000-acre cattle business from his home in southern Montgomery County. In addition to his herd of brood cows, Trotman owned one of the state's largest feeder cattle operations. The 1966 ACA president also served as president of the Montgomery Production Credit Association and as chair of the Intermediate Credit Bank of New Orleans Advisory Committee.[64]

In southwestern Alabama Arthur Tonsmeire, Jr., and M. C. Stallworth, Jr., presented examples of cattlemen with a variety of business interests. Tonsmeire owned large cattle farms in Mobile and Marengo counties while serving as president of First Federal Savings and Loan Association and American Federal Life Insurance Company, both in Mobile. Stallworth, a founding member of the ACA, controlled vast acreages of timber and pasture in southwestern Alabama and in Montgomery County. In addition, his many business interests included a Ford dealership in Mobile and a lumber company in Honduras. In the Tennessee Valley Bo Howard owned a diversified 1,000-acre farm producing cotton, corn, and grains in addition to supporting his herd of more than 300 brood cows. Howard, who had attended Bryson College, was also a member of the Alabama Farm Bureau's executive committee.[65]

By 1960 commercial cattle raising occupied an important position in the agricultural economies of the former plantation districts. The conversion from row cropping to livestock raising combined with the gradual post–World War II mechanization of cotton cultivation and harvesting to bring about a massive outmigration of former black sharecroppers and tenants in the 1950s. The Black Belt sat atop the state's cattle industry and ranked among the South's most productive commercial cattle regions. Some agricultural experts of the early and middle twentieth century claimed the region's dominance in the young industry derived from the superior intelligence and agricultural acumen of the Black Belt's small majority of large landowners. The reasons, however, are numerous. The conversion from cotton to cattle resulted initially from a response to boll-weevil destruction. Several other factors influenced the growth of Black Belt cattle raising as well. Large landholdings of hundreds and thousands of acres provided owners with the potential pastures and forage fields required for commercial cattle herds. Black Belt planters were among the state's few farmers who possessed the capital needed for large outlays of cash up front, and when the 1930s brought credit associations, the region's cattlemen took advantage of this by distributing and obtaining livestock loans. Furthermore, and perhaps most important, Black Belt agriculturists enjoyed the benefits of the information and assistance supplied by government agencies. In this respect these planters and farmers were at least more knowledgeable than their counterparts in other regions, though of course not inherently more intelligent. Through these methods the Black Belt carried its traditional dominance of Alabama agriculture into the cattle industry of the middle twentieth century.[66]

In the decades after 1960 Black Belt cattlemen would be challenged for supremacy in the Alabama cattle industry by an unlikely region. Though a number of restrictions had hindered the expansion of cattle raising into the Appalachian counties before 1960, hill farmers became increasingly involved in cattle farming in the next two decades, eventually rivaling the generations-old cattle ranches of the Black Belt. In the process the industry became increasingly dominated by part-time farmers, revealing the demise of the traditional family farm and the desire by many Alabamians to maintain ties to their agrarian roots.

Chapter 6

New Farmers in the New South

The post–World War II agricultural revolution has produced a number of results. Governmental and technological influences have accomplished a serious reduction of row cropping in Alabama and a significant increase of livestock and poultry raising. In the process the state's farm population, like the nation's, has fallen to minuscule proportions. Within the cattle industry the past three and a half decades have witnessed numerous developments as well. Technological and governmental forces have encouraged the growth of a widespread, multiregional cattle industry in Alabama, one no longer centered in the Black Belt. The steady importation of new, "exotic" breeds has answered the changing demands and preferences for beef and has altered the appearance of cattle herds. The practice of cattle raising has become increasingly dominated by part-time farmers, reflecting considerations of both economics and traditional agrarianism. This development also reflects Alabama's bottom-level position in the cattle industry as a "cow-calf" producer as opposed to one focused on "finishing" cattle in modern feedlots. As the initial actors on the supply side of a relatively unregulated market—especially when compared with most other agricultural commodities—Alabama cattle raisers have also experienced decades of price swings and market reactions quite unfamiliar to other farmers. In recent years cattlemen have also been faced with changing beef appetites resulting from growing health and fitness concerns and with demands and threats from the surging environmental movement.

The cattle industry's relative prominence in Alabama agriculture peaked in the late 1950s before the thriving poultry business took over the top spot. Nevertheless, cattle raising continued to gain popularity in the 1960s and 1970s, and cattle numbers grew steadily before reaching an all-time high in the 1970s. Cattle raising continues to be the most widely diffused and popularly practiced phase of agriculture in the state. This fact is a reflection of Alabama's position as a cow-calf state in the national beef industry, that is, a state in which the primary practice of cattle raisers involves maintaining a herd of brood cows and raising their offspring to a weight of between 400 and 600 pounds before selling them at auction. These calves, known as "feeders" or "stockers," are then purchased by stocker buyers who specialize in feeding cattle until they are sold to the feedlot.

Today the feedlot or finishing center of the United States continues to lie in the Midwest, though there are scattered examples of cattle feeders in Alabama and elsewhere in the South. At one point in the state's history, however, a concerted effort was made to promote cattle feeding or finishing within Alabama and to sever the long-held, dependent connections with midwestern feeders and packers. This effort required not only the establishment of large cattle feeding operations but also the expansion of existing processing capabilities.

In the early years of Alabama's cattle expansion the paucity of meat processors reinforced the state's tendency toward the cow-calf business. Conversely, the lack of feeders and feedlots further slowed the expansion of the meat packing industry. At a 1938 meeting of Birmingham meat packers and dealers, reports revealed that over 75 percent of the city's 500 head per week consumption was shipped in from other states, with three-quarters of the total coming from Texas and Oklahoma. After World War II agricultural experts turned their attentions toward remedying this problem in the state. Despite steady growth, by 1950 only six Alabama meat processors slaughtered at least 5,000 head of cattle annually, and the vast majority of packing operations were small shops serving only local markets. The slaughter cattle industry expanded along with the cattle-raising industry in the 1950s. By 1958, forty-six Alabama packing plants slaughtered over 300,000 pounds of liveweight beef, or approximately 3,000 to 5,000 head or more. Nevertheless, most such establishments continued to be relatively small and locally oriented. The state's meat packers continued to rely on western and midwestern feeders for mature, "fed" cattle. In addition, because freight rates tended to be lower for butchered meat than for live cattle, Alabama's cattle raisers continued to outpace the state's packing industry.[1]

In the late 1950s and early 1960s Alabama agriculturists conducted a determined campaign to establish a substantial cattle feeding industry. The catalysts for this movement emanated from several sources. Progressive agriculturists had long sought diversity, and finishing cattle was one more block of a diversified farming foundation. Alabama packers relished the thought of cheaper, locally grown beef. Perhaps most important, when considering the timing of the movement, heightened interregional tensions created by the civil rights movement were manifested in separatist and independent courses of action among the state's farmers and agribusinessmen. Alabama cattlemen had for decades watched as midwestern feeders purchased their calves and shipped them out of the state for fattening. However, the growing prominence of cattle raising after World War II supplied power to the voices of cattlemen and government agents seeking a diversified cattle industry, and the threats to the traditional structure of southern society lent their message a sense of urgency.

While the demand for grain-fed beef increased dramatically in the 1950s because of the expansion of retail grocery chains and restaurants, Alabamians continued to ship their calves west for finishing. By the early 1960s only one in three cattle marketed in the state was a slaughter animal, or immediately ready for butchering, and many of these were old brood cows and dairy cows. Almost two-thirds of feeder calves marketed in Alabama were shipped out of the state for finishing and over one-half of these had destinations outside the South—and these are likely conservative estimates.

In a 1962 article Auburn livestock marketing specialist Daniel Linton, Jr., estimated that four in five calves produced in Alabama eventually ended up on western and midwestern feedlots. Such statistics spurred the champions of the feeder cattle industry in Alabama.[2]

Cattle feeding was not a new phenomenon in Alabama. For decades some of the state's most prominent cattlemen had "wintered" steers to utilize vast acreages of pasture or surplus grains. In Wilcox County, for instance, O. G. McBeath made use of his Black Belt lands by purchasing steers in the fall at local auctions and selling them in the spring. In the fall of 1954 McBeath bought seventy-five steers at an average price of thirty-three dollars per head. The following spring he marketed the animals at an average of ninety-nine dollars per head. A much larger feeder was E. P. Lane of Montgomery County. As a result of cotton acreage restrictions, Lane turned to producing more corn and oats. This surplus grain and a large, idle labor supply of croppers and tenants influenced Lane's decision to winter more than 1,000 head of cattle per year, which he purchased from auctions throughout central Alabama.[3]

The building of feedlots was quite new, however. The 1950s witnessed the establishment of about 100 large feedlots in the South. One of the earliest in Alabama was built near Decatur in 1959. Operated by Red Hat Feeds, a division of Alabama Flour Mills, and funded also by Armour and Company, the 175-acre feedlot had a capacity of 10,000 head of beef cattle per year. The feedlot owned none of the cattle but instead served as a holding facility. Customers fattening their livestock there paid six cents per head each day and were charged with feeding costs. Other kinds of feedlots existed as well. Some were owned directly by large meat-processing companies, and others purchased cattle to finish and sell to processors.[4]

Cattle finishing increased substantially in Alabama during the early and mid-1960s. By 1965 the state contained thirty-eight commercial feedlots with a capacity of at least 500 head of cattle per year. Between April 1963 and April 1966 the number of Alabama cattle on feed expanded by more than 16 percent to over 40,000 head. Nevertheless, such numbers paled in comparison with midwestern feeder statistics. The element that lay behind the Midwest's predominance in the industry, the production and availability of grains, prevented Alabama from becoming a feeder cattle state. Alabama's production of corn and other feed grains fell far short of meeting feedlot demands. Attempts to overcome the grain shortage produced reduced freight rates on grain imported from the Midwest and the development of "Big John," a new huge, lightweight hopper car used by the Southern Railway, but resulted in no lasting improvements. Furthermore, the increasing prominence of part-time cattlemen reinforced the state's position as a cow-calf industry. Consequently, as the storm of racial and regional strife blew over and as cattle prices rebounded in the late 1960s and early 1970s, the movement to establish a large cattle feeding industry in Alabama waned. In the words of former Alabama Cattlemen's Association executive vice-president E. Ham Wilson, cattle feeding did not "fit into the picture."[5]

At the same time agricultural leaders were promoting cattle feeding in Alabama, new scientific developments and imported breeds increased the efficiency and productive capabilities of cattle raisers. Scientific advances generated by representatives of Auburn University and the agricultural experiment station provided cattlemen with modern, agribusiness methods of production and management. The 1960s also witnessed the beginnings of "exotic" breed importation, as cattlemen gradually abandoned the once exclusive reliance on the most common three breeds—Hereford, Angus, and Shorthorn.

With the eradication of the screw worm in the early 1960s and steady progress in the fight against brucellosis, cattlemen and researchers turned their attentions to production and efficiency. By 1960 pasture management and crossbreeding studies were well under way at Auburn and at numerous experiment substations. In 1961 Auburn University opened its new J. E. Lambert Meats Laboratory, named in honor of the prominent cattleman and alumnus. The university also expanded its annual Beef Cattle Field Day and Bull Sale, which was founded in 1951. In 1964 the extension service and the ACA sponsored the founding of the Beef Cattle Improvement Association. The BCIA, which began work on May 1, 1964, with sixty-seven herds in twenty-nine counties, provided participating cattlemen with methods for keeping better records and for systematically weighing and grading calves. This program reflected the extension service's tradition of encouraging efficiency and standardization. Furthermore, experiment substations hosted numerous field days to discuss practices from winter feeding to insect control. In 1977 the experiment station constructed a new bull test center, emphasizing the importance of over twenty-five years of work in bull testing by Auburn University and the experiment station. That same year the state assembly approved an ACA-sponsored bill appropriating money for the renovation of the livestock judging arena at Auburn, which was eventually renamed the Ham Wilson Livestock Arena.[6]

As Ham Wilson observed, the primary purpose of scientific and technological work was to develop "an ideal type beef cattle." Until the mid-1960s this was accomplished through crossbreeding and food management using the traditional beef breeds such as Hereford and Angus. Auburn's thirteenth annual Beef Cattle Field Day and Bull Sale in 1963 featured only three breeds—Angus, Hereford, and Polled Hereford. In the late 1960s, however, several Alabama cattlemen began importing previously unheard of breeds of cattle to improve the quality of their stock. Like cattle feeding in the same decade, this was not a new phenomenon, but the scale of importation and the effect on the state's cattle population became quite impressive and long lasting.[7]

The first "exotic" imported in significant numbers was an Indian breed, the Brahman. Though there were a few Brahman cattle scattered across the South as early as the antebellum period, the breed did not gain popularity until the 1930s and 1940s. Boosters lauded the peculiar-looking animals as the ideal stock for the South. Like the native cattle of Spanish descent, the Brahman thrived in hot conditions better than the British breeds and was more resistant to disease. Unlike native cattle, Brahmans were hearty animals that fattened quickly when penned and fed, and the Indian breed produced a greater percentage of quality beef than did the popular British breeds.[8]

Before and during World War II cattlemen in the southern half of Alabama began experimenting with Brahmans. Early breeders, such as Franklin Smart of Baldwin County, bred Brahman bulls to native cattle, increasing the average weaning weight of calves by as much as 150 pounds. By World War II Brahman bulls were numerous in the swampy bottoms of southwestern Alabama. Black Belt cattlemen also experimented with the crossbreeding of Brahman and British breeds. Early Black Belt agriculturists who introduced Brahmans to the region included J. Bruce Henderson, G. M. Edwards of Union Springs, Charlie Stallworth of Montgomery County, and Wilson Coker of Lowndes County.[9]

Though the Alabama Brahman Association was only established in 1973, the breed has exercised a much greater influence on the cattle industry than this would suggest. At least four popular new breeds are derived from Brahman-British crossbreeding. The most famous of these, the Santa Gertrudis, was developed on the King Ranch in Texas and imported into Alabama after World War II. Officially three-eighths Brahman and five-eighths Shorthorn, a Santa Gertrudis bull owned by John Armstrong registered the top weight-gaining ability in Auburn's Beef Cattle Performance Testing Program in 1958. Other Brahman-derived breeds that have spread across the state in the past quarter century are the Brangus, three-eighths Brahman and five-eighths Angus; the Braford, three-eighths Brahman and five-eighths Hereford; and the Beefmaster, a combination of Brahman, Hereford, and Shorthorn stock.[10]

By far the most popular "new" breed of the post–World War II era, however, was the Polled Hereford. Developed by an Iowa cattleman who bred the horns out of a select group of Herefords, the Polled Hereford began to replace its horned relative after the war, and Alabama quickly became a southern leader in raising the breed. The Alabama Polled Hereford Association, founded at Demopolis in 1949, consisted of a number of the state's most prominent cattlemen, including president and future ACA president J. E. Lambert. By the 1970s the Polled Hereford had become the dominant breed in the South.[11]

The past three decades have witnessed a steady stream of exotic breed importations into the United States. Cattlemen in Alabama and across the nation have formed associations to promote such breeds as the Chianina, Gelbvieh, Limousin, and Simmental. Easily the most important of the foreign breeds has been the Charolais, a white or straw-colored breed developed in central France. Gaining widespread notoriety only in the 1970s, the Charolais today rivals the Polled Hereford and Angus in popularity among Alabama cattle raisers. The earliest breeder of Charolais cattle in Alabama, and one of the earliest in the nation, was Cecil Shuptrine of Selma. In 1965, two years before helping organize the Alabama Charolais Association, Shuptrine became president of the Eastern Charolais and Charbray Association. In October 1967 he sponsored the first Charolais-Charbray sale in Alabama at Hooper Stock Yards in Montgomery.[12]

Like the Brahman, the Charolais won wide acclaim for its rapid rate of weight gain. As early as 1959 researchers in Louisiana conducted crossbreeding experiments with Charolais and British breeds, and the positive results heightened interest in the French cattle. In 1975 former Auburn beef specialist and executive vice-president of the American International Charolais Association, Dr. J. W. Gossett, declared, "Charolais cattle have revolutionized the U.S. beef industry. During the past twenty years they have created more change and genetic potential than has occurred in the past century." The growing popularity of the Charolais and other breeds served to threaten some raisers of the more traditional British breeds. A 1975 advertisement for a Madison County Polled Hereford farm reacted harshly to the new exotics, touting its stock as truly American cattle and showing them standing red and white in front of a blue sky.[13]

The increasing interest in foreign breeds reflected the growing popularity of commercial cattle raising throughout the state. A leading force behind cattle expansion and a beneficiary of the industry's spread was the Alabama Cattlemen's Association. Under the direction of Ham Wilson, and powered by funds generated by the beef checkoff, the ACA in the 1960s began a cycle of growth that would eventually make it the largest state cattlemen's organization in the country. In 1964, at the ACA's urging, Governor George Wallace proclaimed October "Alabama Beef Month." That same year the ACA moved into its new building in downtown Montgomery and cooperated with the extension service in organizing the Beef Cattle Improvement Association. On the national scene the cattlemen's organization lobbied successfully for congressional passage of a bill placing quotas on beef imports.[14]

After years of phenomenal growth, in 1970 ACA membership reached a total of 12,137, making it the largest such organization in the United States. Four years later membership topped the 18,000 mark, the highest total in the association's history. Beginning in 1974 several years of depressed cattle prices drove many agriculturists out of the cattle business and reduced ACA membership rolls. Nevertheless, the ACA continued to wield considerable political power. In 1975 the ACA secured passage of an Alabama bill prohibiting state agencies from spending tax dollars on imported foreign beef. This was in large part a reaction to the lifting of beef import quotas by the Nixon administration in 1972.[15]

Another development within the ACA reflected the changing demo-

The Alabama Cattlemen's Association's headquarters and "Mooseum" near the capitol in Montgomery. (Courtesy of Alabama Cattlemen's Association)

graphics of cattle raising in Alabama. Before 1970 the center of power in the ACA was located overwhelmingly in the Black Belt. Through that year two-thirds of the association's twenty-five presidents had lived or owned cattle ranches in the Black Belt. After 1970 the ACA's leadership represented a wider spectrum of regions. Of the first fifteen presidents after 1970 only three owned Black Belt cattle farms, and for the first time representatives from Piedmont and Appalachian counties ascended to the organization's top office. Similarly, in 1959 four of the top eight county memberships were found in the Black Belt. In 1975 only one of the seven leading county associations was in the Black Belt. Between 1980 and 1996 the ACA's presidents came from almost every area of the state, from the southwest to the Tennessee Valley and from the Black Belt to the Appalachian counties (Table 4, appendix).[16]

Changes in the ACA revealed a more fundamental shift in the state's cattle industry. After 1960 Alabama agriculture underwent a transformation whereby the commercial cattle-raising industry spread beyond the boundaries of the old plantation districts. In the past three and a half decades the growth of cattle raising has occurred almost exclusively outside the Black Belt. While cattle populations in most Black Belt counties have leveled off or even declined, beef cattle numbers in other regions of the state have shot upward, especially in the Appalachian and Wiregrass counties. This has been the most important development in Alabama cattle raising in the period after 1960 (Table 5, appendix).

Commercial cattle raising played an integral role in the transformation of Alabama agriculture in the quarter century after World War II. Into the 1950s the small farming areas of Appalachia and the Wiregrass were virtually ignored by extension beef specialists, and the bottleneck in cotton harvesting technology—mechanized spindle-type pickers were still rare in the early 1950s—allowed these farmers to harvest small crops of cotton with some degree of profitability. Furthermore, especially in the hill counties of northern Alabama, many small farmers produced Grade B milk for local cheese and ice cream plants. This system unraveled quickly in the late 1950s and 1960s, however, and most farmers who did not leave the land were forced to abandon old ways of agriculture and to adapt to new circumstances. For the majority of farmers beef cattle raising became a key element of survival.

The 1960s brought the widespread use of mechanical cotton pickers to the large and relatively flat farms of the river bottoms and plantation districts and the tightening and reapportioning of federal acreage allotments. Consequently, small cotton farmers who lacked the fertile soils and vast acreages necessary to compete in the rapidly mechanizing field were forced to find off-farm employment, which was plentiful during most of the period, or to alter farming practices. In both cases farmers frequently utilized beef cattle. As a result the small farming areas witnessed a dramatic decline in row cropping and an increase in livestock raising after 1960. In the Wiregrass, for instance, cotton acreage in Geneva County fell from 18,430 in 1959 to 1,086 in 1974. Over the same period the number of cattle and calves more than doubled from 21,115 to 47,238 and hay crops jumped from 693 acres to 8,708. In Houston County cotton acreage plummeted from 20,868 in 1959 to 1,528 in 1974, while the cattle population grew from 23,807 to 48,548.[17]

Even more impressive were changes in the Appalachian counties of northern and northeastern Alabama. In this most agriculturally diversified section of the state, small farmers dotted the landscape on plots of land ranging from a few dozen to a few hundred acres. The small scale of their enterprises made them unattractive to extension agents peddling the beef cattle solution after World War II, and, as the semisubsistence traditions of the past slipped ever further into memory, these farmers explored a number of diverse avenues in order to maintain their lands and improve their livelihoods. In addition to small creek-bottom and hillside plots of cotton, corn, and sorghum, many hill farmers acquired small herds of Jersey or Guernsey cattle to produce Grade A or B milk for the burgeoning dairy market. Others entered the fledgling poultry business. A few larger farmers had the capital and lands needed to take up beef cattle raising. E. N. Vandegrift of Blount County entered the commercial cattle business in 1949 and by 1960 had amassed holdings of 1,650 acres and 175 Polled Hereford brood cows. Such cases were rare in the small farming district, however.[18]

By the late 1950s the mechanization of the cotton harvest and the small hill farmer's inability to compete loomed large on the horizon. Experts estimated that a farmer needed at least 100 acres of cropland in order to profitably utilize a mechanical picker, and the vast majority of hill farmers grew only a small percentage of this. In the following decade stricter sanitary requirements for milk demanded larger-scale dairying or abandonment of the old five- or ten-cow milking operation. Scientifically advanced seeds and fertilizers and improved technology resulted in larger yields on fewer acres and thus federal allotment cutbacks. These factors pushing hill farmers away from row cropping were answered with agricultural and industrial pull factors as well. High cattle and poultry prices in the 1960s made these avenues profitable and desirable, and increased extension service beef cattle activities revealed opportunities and methods outside of row cropping. At the same time growing towns and regional cities offered struggling farm families a way off their land or often a means through which to save their land.[19]

Hill farmers began raising beef cattle through a number of processes.

Some, like Vandegrift, entered the business even before the massive transformation of the 1960s and 1970s. W. J. Bailey of Shelby County bought a few head of beef cattle before World War II and, along with John Rucker, gradually built up a small herd. By 1960 the two men owned a herd of thirty-two registered Polled Hereford brood cows on a 225-acre farm near Montevallo. In Walker County Mayo Guthrie started with four heifers and a bull in 1948. Thirteen years later he grazed forty-four head on 100 acres of improved pasture. Coosa County farmer T. A. McEwen began purchasing Jersey and Holstein calves from local dairy farmers after World War II. In the 1950s he bought a purebred beef bull and cows to mix with his dairy cattle and grazed them on his 200-acre farm and his brother-in-law's 1,800-acre spread. By the late 1950s he owned close to 100 head of cattle and was one of the county's most prominent cattlemen.[20]

In several communities the Soil Conservation Service assisted farmers with the planting of pastures, though much of their energy went into terracing hillsides to reduce row-cropping erosion. In 1967 SCS officials in Cullman County oversaw the planting of 1,800 acres of permanent pasture, bringing the county program's twenty-year total to more than 60,000 acres. One beneficiary of such federal assistance was Cherokee County farmer Sam Burnett. In 1964 he boasted a pasture of fescue and white clover six inches deep and grazed a herd of sixty brood cows on his Boaz farm. Nevertheless, in a bit of irony and a familiar failure to distinguish regional diversity, one week after it offered the Burnett story the *Sand Mountain Reporter* carried an extension service–sponsored article entitled "Despite Troubles Dixie Is Still Land of Cotton." At the very moment the region was in the midst of a necessary move away from cotton planting, the area's small farmers continued to hear praises sung for the white fiber.[21]

Many small farmers initiated beef cattle programs as part of the transition from full-time to part-time farming. Farmers who had maintained small dairy herds frequently kept part of their livestock after leaving the dairy business and gradually added on to the herd by purchasing beef bulls and heifers. In Walker County Bessie Odom and her family converted their small dairy–row crop farm to a beef cattle operation in the 1960s. With fewer farm responsibilities Odom was able to take a job as a cook for extra income. Odom's story reflects a common development in the modern cattle-raising industry, especially among the small farms of the Appalachian counties. When farm owners found themselves unable to support families and maintain their lands or desired the amenities of life afforded only to those with steady cash incomes, they often took full-time or part-time jobs in nearby towns and cities. Those who continued to farm in the evenings and on weekends found beef cattle the perfect commodity. A small herd of beef cattle required far less supervision and management than did a dairy farm or acres of row crops. Only winter feeding and calving season demanded more than cursory supervision, leaving time for work or other pursuits. When prices were good part-time cattlemen could receive handsome returns. Furthermore, cattle raising prevented land idleness and served as a traditional or sentimental link to the past for many farmers and farm descendants. Even farmers who left the land often returned in later years. In the late 1950s Noel and Pauline Blevins left their DeKalb County farm for Chattanooga, where Noel found employment at the Tennessee Stove Works. Upon his retirement in the early 1970s, the two returned to their 300-acre plot in Deerhead Cove and purchased a herd of beef cattle.[22]

Growing cattle numbers in the 1960s and growing interest in beef cattle raising spurred the opening of several new auction yards in the hill counties. These auctions, in turn, boosted cattle raising by offering convenient markets to formerly isolated farmers. In late 1969 Keith Parrish established Cullman Stockyards and marketed 1,000 head of cattle at the first sale. By 1970 hill farmers in northern Alabama enjoyed a number of outlets for their livestock, including auctions at Scottsboro, Guntersville, Jasper, and Albertville. In 1973 the Sand Mountain Livestock Market opened for business at Kilpatrick, and the Fort Payne Stockyard initiated sales the following year. By 1980 the Fort Payne Stockyard ranked third in annual receipts among the state's forty-seven licensed auctions, and stockyards in Ashville, Morris, and Cullman ranked among the state's leaders as well.[23]

The result of these developments in the hills has been not only a transformation in hill country agriculture but also a restructuring of the state's cattle-raising demographics. In the decade after 1964 cotton production declined by 53 percent in the five-county block of Blount, Cullman, Morgan, Marshall, and DeKalb counties. Over the same period the number of cattle and calves grew 82 percent, from 136,195 in 1964 to almost 250,000 ten years later. The number of cattle in Cullman County almost quadrupled between 1959 and 1974, catapulting the county to third place in the category by the latter date. The other four counties all experienced increases of more than 100 percent during the fifteen-year period, as did St. Clair, Clay, Walker, and Winston counties. Over the same span cattle numbers in most Black Belt counties leveled off or even declined. Herds in Dallas, Hale, Montgomery, and Sumter counties decreased slightly, and only in Lowndes County did cattle numbers experience a significant increase (Table 5, appendix). Though as recently as 1968 the top five cattle-producing counties in the state were in the Black Belt, by the early 1970s that region's supremacy in the cattle-raising industry was clearly threatened both by Tennessee Valley and Appalachian counties in the north and by Wiregrass counties in the southeast.[24]

The past two decades have seen a continuation of this trend. By 1982 two of the state's three leading cattle-producing counties were in the Appalachian region. Today, though a few Black Belt counties continue to rank among the state's leaders, the cattle-raising industry obviously reflects a shift to the northern part of the state. The 1992 census revealed that six of the seven leading counties in terms of cattle numbers were in northern Alabama, and five of these are primarily Appalachian counties. Even when considering the northern counties' greater number of dairy farms, the statistics reveal a shift out of the Black Belt. In 1992 Cullman and DeKalb counties led the state in beef brood cow numbers, and northern Alabama claimed six of the nine top counties in that category. Only three Black Belt counties ranked among the top ten in beef cattle income. Furthermore, while cattle numbers in the Black Belt have been reduced to roughly one-half their zenith of the mid-1960s, farmers and cattlemen in several hill counties have steadily expanded their herds over the past decade and more. As Table 5 (appendix) shows, cattle numbers in Cullman and DeKalb counties, the state's leading cattle-producing and farm income–producing counties, had by 1992 even exceeded those of the boom years of the early 1970s.[25]

The last observation leads us toward one of several explanations for this demographic shift. Cullman and DeKalb counties each entered the 1960s with about 3,000 small farmers and tenants, easily more than any other county in the state. After push-and-pull factors drained tenants and many small, inefficient farmers off the land, survivors were left to consolidate landholdings and alter farming practices. Within a decade and a half the number of farms in these hill counties was almost cut in half. Nevertheless, the majority of those staying on the land were small and diversified or part-time farmers. In both cases beef cattle raising made practical and economic sense. Small landowners whose sole or primary income was not derived from the sale of cattle could better weather the violent price swings and rocky bottoms of a relatively unregulated seller's market. When prices were low these agriculturists could better afford to hold their heifer calves off the market than could larger cattlemen who depended more urgently on the income from these sales.[26]

In the Black Belt, however, different historical circumstances prevailed. As the decade of the 1960s opened to the music of the civil rights movement, the familiar cadence and song of the sharecropper slowly receded into memory. Mechanization of the cotton harvest and heightened awareness of social and racial injustice rendered sharecropping untenable. With the passing of the century-old labor institution, Black Belt landlords and cattlemen lost the majority of their year-round labor supply. Accompanying this was a continued shift in cotton production to the Tennessee Valley. As a result many Black Belt cattlemen came to rely more and more heavily on cattle income. The price crash of the mid-1970s injured many Black Belt cattlemen beyond repair and forced several out of the business. Though Lowndes, Marengo, and Montgomery counties continue to rank among the state's leaders in beef cattle production and income, other Black Belt counties did not recover after 1974. Bullock and Greene counties had descended into obscurity as cattle-raising counties by the 1990s, and cattle numbers in Hale, Perry, and Sumter counties have steadily declined over the past two decades.[27]

A quick examination of the 1992 census reflects persistent differences in farm sizes and numbers between the two regions. In this latest census Cullman and DeKalb counties led the state with respectively 1,490 and 1,378 farms reporting an inventory of cattle and calves. Black Belt cattle farm numbers ranged from 174 in Wilcox County to 454 in Montgomery. As the figures in Table 6 (appendix) indicate, hill counties tended to have significantly fewer cattle per farm than the Black Belt counties. Fewer than one in four farmers in Cullman, DeKalb, Walker, and Jackson counties owned more than fifty head, while over half those in Lowndes County owned fifty or more. More than 40 percent of cattlemen in Dallas, Montgomery, and Wilcox counties owned at least fifty head. The Black Belt also possessed a much larger percentage of very large cattle farms. While the state average for cattle farms in excess of 200 head was only 4.4 percent of farms per county, and in most hill counties the number was less than 3 percent, the percentage of large cattle ranches in most Black Belt counties exceeded 10 percent. Lowndes County contained forty-five such farms (17.2 percent), and twenty-one of these counted more than 500 head. Cattle farms of 200 head or more accounted for one in seven in both Dallas and Marengo counties.[28]

Such statistics reflect not only the dominance of part-time, small farmers in the Appalachian region but also the continuing polarization of the cattle-raising industry. Agriculturists relying primarily on the income from cattle sales have been forced to expand their operations into large ranches because of price swings, pasture demands, and high feed prices. Between these few individuals and the bevy of small, part-time farmers at the other end of the spectrum, only a few small and medium farmers exist, and these are generally well diversified. In the hills many of these farmers are involved in the poultry business or in hog production, while in other regions diversified farmers rely on commodities from peaches to peanuts to catfish.

This unique situation is a result in large part of the relatively unregulated market for cattle raisers. Unlike the cases of most other agricultural commodities, the commercial cattle industry has historically been free of direct federal subsidization. Since the 1930s cattlemen "have prided themselves" on not asking for government subsidies and have consistently associated minimum price laws with the possibility of price controls. Opposition to price legislation during the boom days of the World War II era was particularly vehement. The letters that summoned the delegates to the founding meeting of the Alabama Cattlemen's Association in 1944 called on Alabama's cattlemen to "fight against roll-back in prices on our products and the subsidies that we are constantly threatened with." Not surprisingly, cattle prices in 1943 had been the highest to that date. Similarly, seven years later, with cattle prices at their highest point before the 1970s, the ACA Board of Directors reacted harshly to proposed price controls on meat announced by the National Price Administration. Calling price controls the "tools of socialism which destroy sound economy . . . [and] the liberty of the people," the Board praised Congress for "slapping down" such measures.[29]

Because of the effects of market forces on cattle prices paid at auction, the state's cattle population has responded to national forces of supply and demand. In times of low prices cattle numbers have risen as cattlemen have kept much of their livestock off the market. For instance, Alabama's cattle population reached an all-time high in 1976 of 2.85 million after a 47-percent price slump over a two-year period. However, after prices rebounded and hit an average yearly high of 62.9 cents per pound in 1979, sellers rushed the market and the cattle population plummeted by more than a million head to 1.73 million in 1980. Alabama's cattle inventory reached a forty-year low in 1991 as the average price soared to a record high of 76.6 cents per pound.[30]

Though cattle prices have experienced a cycle of highs and lows since World War II, fixed feed costs have steadily risen. Subsidies on corn, oats, and other grains have maintained relatively high feed costs even during times of low cattle prices. Consequently, most cattle raisers have attempted to grow the bulk of their winter feed. Although this comprised corn, hay, and other grains in the early years of the industry, since World War II Alabama's cattlemen have consistently turned to the production of hay. Yearly harvests of clovers, alfalfa, fescue, and other grasses supply most cattlemen with their winter feed.

In many instances in which cattle replaced cotton, the hay harvest in early years served as an agricultural, economic, and social replacement for the cotton harvest. Though generally quicker and more intense—depending on acreages and yields hay harvests could take two days or two weeks—hay harvests provided a spring or summer occasion for the gathering of family and neighbors in a common pursuit. Old men sat reminiscing under shade trees and directing the goings on while sons and grandsons, and often daughters and granddaughters, hoisted bales or stacked pitchfork loads of hay on wagons. Early in the morning the women hurried to start preparing the big dinner at which the participants would partake of the familiar food and conversation of the hay harvest. The hay harvest also provided cash jobs for tenants, poor farmers, and young men, as cotton picking had done.

In the years since World War II, however, cattle raisers have adopted a

Stacking sorghum hay on a DeKalb County farm, c. 1920s. (Courtesy of ACES Photo Collection, Auburn University Archives)

steady stream of hay-harvesting machines designed to save labor and thus the wages of laborers. In the process the hay harvest described above has become virtually obsolete; a couple of workers using labor-saving harvesting machinery can now do the work that previously required eight or more laborers. Until World War II almost all the hay harvested in the country was loosely stacked in barn lofts or baled into wire-tied, rectangular bales weighing 50 to 100 pounds. The balers in use were stationary and powered by mules, horses, or small engines mounted atop the machines. Farmers carried the hay on wagons or trucks to the edge of the field for baling.

During World War II implement companies began marketing tractor-drawn, self-tying, pick-up balers operated by tractor power takeoffs or mounted engines. In late 1944 International Harvester introduced its new labor-saving twine baler, the Model 50-T. The cheaper sea grass twine allowed for easier handling of bales and grew in popularity as other manufacturers moved to produce twine balers. Some companies sought to capture a share of the market through novelty and creativity; for instance, in late 1947 Allis-Chalmers presented its Roto-Baler, which rolled hay into twine-tied, tubular packages.[31]

As more and more farmers purchased tractors and entered the cattle-raising business, implement makers developed smaller, less expensive balers. The New Holland Machine Company achieved success with such a product in 1953 when it introduced the Model 66, a shorter, lower baler that could be pulled by a one-plow tractor. By 1963 thirteen implement manufacturers were producing a total of forty-four different pick-up balers; John Deere alone had eleven models on the market. Between World War II and 1964 the number of pick-up balers on Alabama farms increased from practically none to more than 6,000. With the baler's basic design established, manufacturers next attacked the labor-intensive jobs of loading bales onto trucks or wagons, transporting the bales to a barn or shed, and stacking the bales. Several versions of the field elevator allowed for easier loading of the bales in the field. Connected to a hay trailer or flatbed truck, the elevator scooped up bales as the driver navigated the field and carried them to a waiting hand who stacked the bales without leaving his position. Even more labor saving was a device such as the Easy Way Baler Lodr made by the Weiss Implement Company of Wisconsin. The Easy Way Baler Lodr provided a chute for John Deere balers that delivered each bale directly to a trailer attached to the rear of the baler. New Holland then developed the Bale Thrower to eliminate the need for a man on the wagon; this machine tossed fresh bales into a cotton-type wagon.[32]

Not satisfied with eliminating field labor needs, implement companies also addressed the problem of stacking hay in barns. Hay elevators, or conveyors, were designed to transport the bales from ground to loft, thus saving the back-breaking chore of tossing seventy-five-pound bales overhead. Many farmers replaced obsolete loft barns with pole barns and large, open sheds to achieve easier hay storage. Such barns were necessary to take advantage of the ultimate labor-saving devices, the automatic bale wagon and the bale accumulator. The Farmhand Company promised "no hand labor . . . from bale to bunk" with its Bale Accumulator, which collected and arranged eight bales at a time straight off the baler and deposited its neat stack on a custom-made bale wagon. Even more advanced was the New Holland Stackliner automatic bale wagon, which not only loaded the hay in the field but also unloaded it in the barn; this machine enabled a farmer to harvest the crop without ever touching a bale of hay.[33]

The introduction of the round baler in the mid-1970s offered the small and part-time farmer the supreme attributes of labor-saving technology and convenience. After attempts by John Deere, Heston, and other manufacturers to market machines that loosely stacked large amounts of hay, implement companies presented the round baler, which twine-ties several hundred pounds of rolled hay into a bale ranging from three by four feet to five by six feet. Requiring only a tractor with a hay fork for transporting the bale, the new round bales allowed small farmers to harvest their hay alone. Furthermore, the massive size of the bales made it possible for cattlemen to avoid daily winter feedings as their herds required more time to finish each large bale. Over the past two decades the steady expansion of round baling has largely eliminated the square baler and the once common family and community efforts during the hay harvest season.[34]

Like the hay harvest, Alabama's cattle industry has changed in the past generation. The Appalachian and Tennessee Valley regions of the north have largely surpassed the once dominant Black Belt in cattle production. Nevertheless, the Black Belt continues to supply a disproportionately high percentage of the ACA leadership, reflecting both the region's traditional prominence in the organization and the region's inordinate number of large, prosperous ranches.[35] Cattle raising has also come to be dominated by small, part-time farmers; in this way the commercial cattle-raising industry has helped maintain the bond between many farm families and the land.

In addition to the cyclical nature of prices, modern cattlemen and their associations are also faced with increasingly harsh attacks and demands. Two of the keystones of modern America—environmentalism and health consciousness—have achieved increasing levels of influence on cattlemen's concerns in the recent past. Over the past decade criticism of the cattle industry has become particularly widespread and media savvy. Groups such as People for the Ethical Treatment of Animals (PETA) and the Farm Animal Reform Movement (FARM) have thrust anti-beef campaigns into the national spotlight. In 1990 singer k. d. lang appeared on the television program "Entertainment Tonight" to tout a public service campaign, "Meat

Stinks." The campaign sought to counter the beef advertisements of the past two decades featuring such personalities as Dinah Shore and James Garner. Environmentalists also used Earth Day 1990 as a forum to deliver the message that people could "help save the earth" by eating less beef. Advertisements sporting this slogan, which appeared in *USA Today* and other widely read newspapers and magazines, noted the industry's waste of water and grain, pollution of water, destruction of habitats, and production of ozone-depleting methane. The publication in 1992 of Jeremy Rifkin's *Beyond Beef: The Rise and Fall of the Cattle Culture*, a scathing denunciation of the subject's dangers to the environment, caused further consternation among cattle raisers and their representatives. Although the greatest part of environmentalist criticism has been leveled against the unsanitary, polluted feedlots of the Midwest and the practices of western cattlemen ranging their herds on government land, blanket accusations have alarmed industry representatives throughout the country and have elicited defensive reactions from many. In 1992 Ham Wilson called the environmentalist onslaught "probably the biggest blow that we have ever had." Congress and the Environmental Protection Agency have also raised questions concerning private property rights and environmental issues with wetlands legislation and the Clean Water Act.[36]

Another popular American issue has affected decisions and actions within the cattle industry. The health consciousness of a growing segment of the American population has influenced meat processors to seek leaner stock, thus leading cattlemen "to breed a different type of cattle." Ironically, in the past few years the Texas Longhorn, whose back fat is one-third to one-half less than that of popular breeds, has undergone a small resurgence. This same lean animal had lost favor after the Civil War as eastern tastes for marbled beef demanded British breeds. Scientists have also attempted to engineer leaner, sleeker genetics within the heavily utilized breeds, and in 1990 Auburn researcher Dr. Dale Huffman developed AU Lean, a low-fat ground beef product that inspired the introduction by McDonald's of the "McLean" burger.[37]

Declining prices since 1992 have compounded cattle industry anxieties. Nevertheless, a national beef checkoff program, passed by Congress in 1985, has increased funds for beef promotion. Furthermore, within Alabama the ACA remains a potent political force under the direction of William E.

The modern cattleman's horse: Don Freeman tends to his crossbred cattle on his Lowndes County farm. (Courtesy of Alabama Cattlemen's Association)

Powell III, holder of an Auburn animal science doctoral degree, who succeeded Ham Wilson as executive vice-president in 1985. In large part because of efforts by Powell and the ACA, in 1995 S & C Beef Processors opened a large beef pattie plant in Montgomery to service the Southeast's fast food market. The cattle industry continues to generate millions of dollars in Alabama and remains the most egalitarian and popular sphere of agriculture. After becoming the state's number one agricultural commodity in the late 1950s, the commercial cattle industry was soon surpassed by the booming poultry business, which maintains its top spot today. In 1993 cash receipts from the sale of cattle and calves amounted to $374.4 million, second only to those of the poultry industry among Alabama's agricultural commodities.[38]

Cattle raising in Alabama, which once flourished among the open-range herders of the piney woods and the mountains, now bears little resemblance to the practice that survived for more than two centuries. The midwestern model adopted by Black Belt planters before World War I and its British breeds have become the standards of the industry throughout the state. In the early twentieth century the Black Belt supplanted the southern piney woods as Alabama's premier cattle region; over the past three decades this position has been challenged by the expansion of cattle raising among the small and part-time farmers of the Appalachian and Wiregrass regions. In the process of growth the cattle-raising industry has played a key role in the state's fundamental agricultural and social transformations. Ironically, the practice of cattle raising, once dominated by plantation-belt planters who sought to eliminate small, inefficient producers, has become a mainstay of the small and part-time farmer clinging to the land and the agrarian tradition.

APPENDIX

Table 1. Alabama Cattle and Human Populations, 1840–1990

Year	Cattle Population	Human Population
1840	668,018	590,756
1850	728,015	771,623
1860	773,396	964,201
1870	487,163	996,992
1880	751,190	1,262,505
1890	875,976	1,513,401
1900	586,337	1,828,697
1910	816,941	2,138,093
1920	1,044,008	2,348,174
1930	681,298	2,646,248
1940	889,983	2,832,961
1950	1,269,389	3,061,743
1960	1,526,410	3,266,740
1970	1,781,061	3,444,165
1980	1,652,824	3,893,888
1990	1,453,137	4,040,587

Source: Donald B. Dodd, *Historical Statistics of the States of the United States: Two Centuries of the Census, 1790–1990* (Westport, Connecticut: Greenwood Press, 1993).

APPENDIX

Table 2. Effects of the Boll Weevil in the Black Belt, 1910–1920

County	Cotton Acres		Cattle	
	1910	1920	1910	1920
Bullock	107,099	47,287	15,428	20,496
Dallas	153,473	98,379	24,684	31,907
Greene	72,751	36,028	16,070	16,505
Hale	88,081	42,702	16,514	22,165
Lowndes	122,629	50,308	21,773	27,739
Marengo	117,257	58,456	29,546	38,044
Montgomery	157,001	48,625	22,572	25,875
Perry	95,757	44,553	16,204	19,711
Sumter	80,494	52,048	21,325	26,819
Wilcox	107,480	52,498	27,945	33,054

Source: U.S. Department of Commerce, Bureau of the Census, *U.S. Census of Agriculture,* 1910 and 1920 (Washington, D.C., Government Printing Office).

Table 3. Cattle and Cotton Prices, Selected Years

Year	Cattle ($/cwt)	Cotton ($/lb)
1910	3.25	.140
1919	6.70	.349
1924	3.20	.231
1929	6.20	.166
1933	2.40	.106
1941	6.40	.174
1951	24.60	.380
1956	11.50	.327
1965	16.40	.301
1973	41.60	.440
1976	28.60	.660
1983	47.00	.658
1991	76.60	.566
1996	41.30	.714

Source: Alabama Agricultural Statistics Service, Montgomery.

APPENDIX

Table 4. Presidents of the Alabama Cattlemen's Association

Year	President	Home County
1944	J. Bruce Henderson	Wilcox
1945	J. Bruce Henderson	Wilcox
1946	Robert J. Lowe	Madison
1947	W. P. Breen	Greene
1948	W. P. Breen	Greene
1949	Will Howard Smith	Autauga
1950	O. J. Henley	Tuscaloosa
1951	J. Ernest Lambert	Wilcox
1952	Mack Maples	Limestone
1953	Preston Clayton	Barbour
1954	M. C. Stallworth, Jr.	Washington
1955	T. Whit Athey, Jr.	Montgomery
1956	Carl B. Thomas	Madison
1957	Mortimer H. Jordan IV	Jefferson
1958	James L. Adams	Houston
1959	Arthur Tonsmeire, Jr.	Mobile
1960	Edward Wadsworth	Autauga
1961	J. Ed Horton, Jr.	Limestone
1962	Richard Arrington	Montgomery
1963	E. R. Howard	Madison
1964	Selden Sheffield	Hale
1965	Richard Beard, Sr.	Jefferson
1966	John M. Trotman	Montgomery
1967	Dr. A. C. Newman, Jr.	Lee
1968	Billy Robertson	Hale
1969	W. Comer Sims	Dallas
1970	Cecil Lane	Lowndes
1971	Harold Johnson	Tallapoosa
1972	Henry B. Gray III	Barbour
1973	W. M. Brown	Escambia
1974	Forrest Killough	Talladega
1975	K. Stanley Drake	Lee
1976	Raymond B. Jones	Madison
1977	Harold E. Pate	Lowndes
1978	W. R. Lanier	Choctaw
1979	Milton Wendland	Autauga
1980	Julio Corte, Jr.	Baldwin
1981	Joe R. Crawford	Morgan
1982	Dr. Billy Powell	Washington
1983	Dr. George S. Killian	DeKalb
1984	James E. Hart, Jr.	Escambia

Table 4. (continued)

Year	Name	County
1985	Dr. George C. Smith	Clay
1986	Steve Tondera	Madison
1987	Ronnie Holladay	Lowndes
1988	A. W. Compton	Marengo
1989	Billy Maples	Limestone
1990	Bill Johnson	Montgomery
1991	Glynn Debter	Marshall
1992	Ned Ellis	Lowndes
1993	Ronny Donaldson	Cullman
1994	Phil Hardee	Monroe
1995	Tim Coe	Randolph
1996	L. D. Fitzpatrick	Lowndes
1997	Bob Helms	Coffee
1998	Greg Blythe	Morgan

APPENDIX

Table 5. Cattle Numbers in Selected Counties, Listed by Region, 1959–1992

County	1959	1974	1992
Appalachia			
Blount	17,367	45,209	39,441
Clay	9,487	19,069	16,811
Cullman	17,027	66,806	67,839
DeKalb	19,483	47,402	55,871
Jackson	20,643	39,480	34,876
Marshall	12,648	42,189	41,053
Tennessee Valley			
Lauderdale	22,678	49,006	39,858
Lawrence	21,373	50,584	30,278
Limestone	26,575	38,149	25,440
Madison	35,977	54,755	27,447
Black Belt			
Greene	32,991	36,511	16,244
Hale	46,837	41,047	30,996
Lowndes	60,406	71,193	38,359
Marengo	54,493	58,352	37,078
Montgomery	89,123	88,975	41,671
Sumter	47,341	45,227	23,980
Southwest			
Baldwin	39,516	50,886	31,358
Clarke	16,797	16,702	7,186
Escambia	16,227	23,131	12,003
Washington	14,242	15,634	9,711
Wiregrass			
Coffee	18,197	31,641	25,751
Covington	26,267	42,085	31,167
Geneva	21,115	47,238	29,267
Houston	23,807	48,548	28,245

Source: U.S. Department of Commerce, Bureau of the Census, *U.S. Census of Agriculture,* 1959, 1974, and 1992 (Washington, D.C., Government Printing Office).

APPENDIX

Table 6. Cattle-Raising Statistics, Selected Counties, 1992

County	Farms with Cattle	Farms with <50	Farms with >200	Farms with >500	Number of Milk Cows
Appalachia					
Blount	829	587	19	5	1,270
Clay	320	211	11	1	392
Cullman	1,490	1,117	39	6	2,020
DeKalb	1,378	1,041	28	1	1,800
Jackson	807	635	22	3	1,294
Marshall	971	718	20	1	1,478
Tennessee Valley					
Lauderdale	869	681	32	3	625
Lawrence	682	500	19	3	851
Limestone	608	476	18	2	547
Madison	529	390	20	5	913
Black Belt					
Dallas	301	158	43	11	687
Hale	276	163	32	15	1,835
Lowndes	261	125	45	21	384
Marengo	350	192	51	11	1,814
Montgomery	454	254	49	13	793
Wilcox	174	100	21	5	44
Southwest					
Baldwin	444	276	36	7	960
Clarke	156	115	6	0	28
Escambia	212	147	10	0	398
Washington	246	193	5	0	51
Wiregrass					
Coffee	425	260	22	1	769
Covington	557	364	19	6	743
Geneva	470	268	19	4	1,085
Houston	417	247	26	5	831

Source: U.S. Department of Commerce, Bureau of the Census, *U.S. Census of Agriculture,* 1992 (Washington, D.C., Government Printing Office).

NOTES

A.A.E.S.   Alabama Agricultural Experiment Station (Auburn: Alabama Polytechnic Institute).
ACA   Alabama Cattlemen's Association.
ACES   Alabama Cooperative Extension Service Records, Auburn University Archives.
AMS Census   Agricultural manuscript census schedules.
API   Alabama Polytechnic Institute, Auburn.
Census Bureau   U.S. Department of Commerce, Bureau of the Census (Washington, D.C., Government Printing Office).
F & I   *Montgomery Advertiser,* "Farm and Immigration" section.
GPO   Government Printing Office, Washington, D.C.
OR   *War of the Rebellion: A Compilation of the Official Records of the Union and Confederate Armies,* 70 vols. in 127 books and index (Washington, D.C.: Government Printing Office, 1880–1901).

1. The Melding of Traditions

1. Terry G. Jordan, *North American Cattle-Ranching Frontiers: Origins, Diffusion, and Differentiation,* Histories of the American Frontier (Albuquerque: University of New Mexico Press, 1993), 106; Joe A. Akerman, Jr., *Florida Cowman: A History of Florida Cattle Raising* (Kissimmee: Florida Cattlemen's Association, 1976), 1, 2; Charles W. Arnade, "Cattle Raising in Spanish Florida, 1513–1763," *Agricultural History* 35 (July 1961): 117.

2. Herbert Ingram Priestley, ed. and trans., "The Luna Papers: Documents Relating to the Expedition of Don Tristan De Luna Y Arellano for the Conquest of La Florida in 1559–1661," *Publications of the Florida State Historical Society* 1, no. 8 (1928): xxxiv, xliii, 63, 101, 103.

3. Akerman, *Florida Cowman,* 3; Jordan, *North American Cattle-Ranching Frontiers,* 106.

4. Jordan, *North American Cattle-Ranching Frontiers,* 106–8; Arnade, "Cattle Raising in Spanish Florida," 119; Akerman, *Florida Cowman,* 18.

5. Jordan, *North American Cattle-Ranching Frontiers,* 120; T. H. Ball, *A Glance into the Great South-East, or, Clarke County, Alabama, and Its Surroundings, from 1540 to 1877* (Chicago: Knight & Leonard, 1879), 40; M. Penicaut, *Annals of Louisiana from the Establishment of the First Colony under M. D'Iberville, to the Departure of the Author to France, in 1722,* trans. B. F. French, Historical Collections of Louisiana and Florida (New York: J. Sabin & Sons, 1869), 98; Jay Higginbotham, *Old Mobile: Fort Louis de la Louisiane, 1702–1711* (Mobile:

Museum of the City of Mobile, 1977), 442, 364; Peter J. Hamilton, *Colonial Mobile* (Boston: Houghton Mifflin Co., 1897; Tuscaloosa: University of Alabama Press, 1976 [Southern Historical Publication no. 20]), 52.

6. Penicaut, *Annals of Louisiana*, 114; Dunbar Rowland and A. G. Sanders, comps., eds., and trans., *French Dominion, 1704–1743*, vol. 3 of *Mississippi Provincial Archives* (Jackson: Press of the Mississippi Department of Archives and History, 1932), 268.

7. Rowland and Sanders, *French Dominion, 1704–1743*, 271; Jordan, *North American Cattle-Ranching Frontiers*, 121; Peter J. Hamilton, *The Founding of Mobile, 1702–1718* (Mobile: Commercial Printing Co., 1911), 57.

8. Penicaut, *Annals of Louisiana*, 101.

9. Jordan, *North American Cattle-Ranching Frontiers*, 109–10; John Solomon Otto, *The Southern Frontiers, 1607–1860: The Agricultural Evolution of the Colonial and Antebellum South* (New York: Greenwood Press, 1989), 30–39.

10. Jordan, *North American Cattle-Ranching Frontiers*, 178; Hamilton, *Colonial Mobile*, 251, 255, 258; Robin F. A. Fabel, *The Economy of British West Florida, 1763–1783* (Tuscaloosa: University of Alabama Press, 1988), 112; Robert R. Rea, "Planters and Plantations in British West Florida," *Alabama Review* 29 (July 1976): 229, 232.

11. Mrs. Dunbar Rowland, "Peter Chester, Third Governor of the Province of West Florida under British Dominion, 1770–1781," *Publications of the Mississippi Historical Society* 5 (1925): 114, 125, 120.

12. Michael F. Doran, "Antebellum Cattle Herding in the Indian Territory," *Geographical Review* 66 (January 1976): 49.

13. Benjamin Hawkins, *A Combination of a Sketch of the Creek Country in the Years 1798 and 1799 by Benjamin Hawkins and Letters of Benjamin Hawkins, 1796–1806*, vol. 3, pt. 1 in *Collections of the Georgia Historical Society* (Savannah: Georgia Historical Society, 1848; Spartanburg, South Carolina: Reprint Company, 1982), 19, 23, 30, 41, 45, 38.

14. Ibid., 26, 38, 44; George E. Brewer, "History of Coosa County," *Alabama Historical Quarterly* 4 (Spring 1942): 126; Dan McGillivray to Panton, 5 June 1799, Papers of Panton, Leslie & Co., collection at the University of West Florida, Pensacola (microfilm, Auburn University); Ball, *Glance into the Great South-East*, 61.

15. Benjamin Hawkins to Th. Jefferson, 23 January 1800; Hawkins to John Milledge, 11 August 1805; Tuskeegee Tustunnuggee to Hawkins, 21 December 1808; Chiefs of the Lower Towns to Hawkins, 14 March 1809, box 1, Benjamin Hawkins Papers, Auburn University Archives (photocopies).

16. Hawkins to Governor of Georgia, 7 July 1813; Hawkins to Capt. Carr, 2 August 1813; Hawkins to Maj. Gen. Jackson, 27 February 1815, ibid.

17. Jordan, *North American Cattle-Ranching Frontiers*, 178; Ball, *Glance into the Great South-East*, 51.

18. Mrs. Dunbar Rowland, "Peter Chester," 60, 88.

19. Ibid., 173; Bernard Romans, *A Concise Natural History of East and West Florida* (New York, 1775; New Orleans: Pelican Publishing, 1961), 119–20, 125.

20. William Bartram, *Travels through North & South Carolina, Georgia, East & West Florida, the Cherokee Country, the Extensive Territories of the Muscogulges, or Creek Confederacy, and the Country of the Chactaws* (Philadelphia: James & Johnson, 1791), reprinted as

William Bartram Travels with an introduction by Robert McCracken Peck (Salt Lake City: Peregrine Smith, 1980), 196–97, 263, 278.

21. John Pope, *A Tour through the Southern and Western Territories of the United States of North America* (1792; reproduced with an introduction and indexes by J. Barton Starr in the Bicentennial Floridiana Facsimile Series, Gainesville: University of Florida Press, 1979), 49, 64.

22. John Melish, *Travels through the United States of America in the Years 1806 & 1807, and 1809, 1810, & 1811* (London: George Cowie and Co., 1818; New York: Johnson Reprint Corp., 1970), 391–92; Walter Brownlow Posey, ed., "Alabama in the 1830's; As Recorded by British Travelers," in *Birmingham-Southern College Bulletin* 31 (December 1938): 5, 6, 31, 33.

23. Otto, *Southern Frontiers*, 83, 85.

24. Washington County, Mississippi Territory Tax Rolls, reel 13, Mississippi Research Microfilm, William B. Hamilton Papers, Duke University Special Collections, Durham, North Carolina.

25. Archer B. Hulbert, *The Paths of Inland Commerce: A Chronicle of Trail, Road, and Waterway,* The Chronicles of America Series (New Haven: Yale University Press, 1920), 22–23.

26. H. Toulmin, "A Geographical and Statistical Sketch of the District of Mobile," *American Register* 6 (1810): 334, 336, 338, 339.

27. Richard S. Lackey and John D. W. Guice, *Frontier Claims in the Lower South* (New Orleans: Polyanthos, 1977), xiv, xv, 31, 63.

2. Piney Woods and Plantations

1. See Frank L. Owsley, "The Pattern of Migration and Settlement on the Southern Frontier," *Journal of Southern History* 11 (May 1945): 147–76; John D. W. Guice, "Cattle Raisers of the Old Southwest: A Reinterpretation," *Western Historical Quarterly* 8 (April 1977): 166–87; Terry G. Jordan, *Trails to Texas: Southern Roots of Western Cattle Ranching* (Lincoln: University of Nebraska Press, 1981); John Solomon Otto, "The Migration of the Southern Plain Folk: An Interdisciplinary Synthesis," *Journal of Southern History* 51 (May 1985): 183–200; and John Solomon Otto, *The Southern Frontiers, 1607–1860: The Agricultural Evolution of the Colonial and Antebellum South,* Contributions in American History, no. 133 (New York: Greenwood Press, 1989), 3, 4.

2. Forrest McDonald and Grady McWhiney, "The Antebellum Southern Herdsman: A Reinterpretation," *Journal of Southern History* 41 (May 1975): 147–66; McWhiney, "The Revolution in Nineteenth-Century Alabama Agriculture," *Alabama Review* 31 (January 1978); Eugene D. Genovese, "Livestock in the Slave Economy of the Old South: A Revised View," *Agricultural History* 36 (July 1962): 143–49.

3. McDonald and McWhiney, "Antebellum Southern Herdsman," 155.

4. William Warren Rogers, Robert David Ward, Leah Rawls Atkins, and Wayne Flynt, *Alabama: The History of a Deep South State* (Tuscaloosa: University of Alabama Press, 1994), 53; James Stuart, *Three Years in North America* (Edinburgh: Robert Cadell, 1833), 142, 150.

5. William Warren Rogers et al., *Alabama,* 90–91.

6. Lewis Cecil Gray, *History of Agriculture in the Southern United States to 1860,* 2 vols.

(Gloucester, Massachusetts: Carnegie Institution of Washington, 1958), 894; Owsley, "Pattern of Migration," 159, 161; Thomas D. Clark and John D. W. Guice, *Frontiers in Conflict: The Old Southwest, 1795–1830,* Histories of the American Frontier (Albuquerque: University of New Mexico Press, 1989), 105, 106.

7. Adam Hodgson, *Letters from North America Written during a Tour in the United States and Canada,* vol. 1 (London: Hurst, Robinson, & Co., 1824), 148, 149; T. H. Ball, *A Glance into the Great South-East, or, Clarke County, Alabama, and Its Surroundings, from 1540 to 1877* (Chicago: Knight & Leonard, 1879; Grove Hill, Alabama, 1882), 186, 325; Benjamin Franklin Riley, *History of Conecuh County, Alabama* (original printing 1881; Blue Hill, Maine: Weekly Packet, 1964), 217.

8. Wendell H. Stepp and Pamela Ann Stepp, *Dothan: A Pictorial History* (Norfolk, Virginia: Donning, 1984), 10; F. L. Cherry, "The History of Opelika and Her Agricultural Tributary Territory," *Alabama Historical Quarterly* 15 (1954): 205, 203, originally published in the *Opelika Times* between 5 October 1883 and 17 April 1885; Fred S. Watson, *Coffee Grounds: A History of Coffee County, Alabama, 1841–1970* (Anniston, Alabama: Higginbotham, 1970), 18, 23.

9. Philip Henry Gosse, *Letters from Alabama* (London: Morgan and Chase, 1859), 271.

10. Statistics gathered from 1850 agricultural manuscript census schedules of Covington and Washington counties.

11. Covington and Washington counties, AMS Census 1850.

12. J. F. H. Claiborne, "A Trip through the Piney Woods," *Publications of the Mississippi Historical Society* 9 (1906): 514–16.

13. Ibid., 521–22.

14. Sam Bowers Hilliard, *Hog Meat and Hoecake: Food Supply in the Old South, 1840–1860* (Carbondale: Southern Illinois University Press, 1972), 120.

15. Paul Wayne Taylor, "Mobile: 1818–1859 as Her Newspapers Pictured Her," M.A. Thesis, University of Alabama, 1951, 43–44.

16. John Knox, *A History of Morgan County, Alabama* (Decatur, Alabama: Morgan County Board of Revenue and Control, 1967), 15; Hodgson, *Letters from North America,* 271; W. F. Withers to C. C. Clay, 21 August 1839, box 2, file 1, Clement Claiborne Clay Papers, Duke University Special Collections, Durham, North Carolina.

17. DeKalb and Randolph counties, AMS Census 1850 and 1860.

18. John Perry Cochran, "James Asbury Tait and His Plantations," M.A. Thesis, University of Alabama, 1951, 90–91.

19. Record Book, John Park Papers, Alabama Department of Archives and History, Montgomery; *Report of the Commissioner of Patents for the Year 1848* (Washington, D.C., 1849), 504–5; Weymouth T. Jordan, *Hugh Davis and His Alabama Plantation* (University, Alabama: University of Alabama Press, 1948), 125–26.

20. James Benson Sellers, *Slavery in Alabama* (University, Alabama: University of Alabama Press, 1950), 28, 30; Lauderdale County, AMS Census 1860.

21. J. H. Dent Farm Journals, Account Books, vol. 3, p. 177 (microfilm, Auburn University); Memoranda (James A. Tait), 8 May 1825, box 1, folder 5, Tait Collection, Auburn University Archives; Robert William Fogel and Stanley L. Engerman, *Time on the Cross: The Economics of American Negro Slavery* (Boston: Little, Brown & Co., 1974), 42.

22. John Horry Dent Farm Journals and Account Books, box 1, folder 1 (Farm Journals, 1840–1842), Auburn University Archives (photocopied collection), original at University of Alabama.

23. James C. Bonner, "Profile of a Late Ante-Bellum Community," *American Historical Review* 49 (July 1944): 675; Gray, *History of Agriculture,* 837; Hilliard, *Hog Meat and Hoe Cake.*

24. Charles T. Leavitt, "Attempts to Improve Cattle Breeds in the United States, 1790–1860," *Agricultural History* 7 (April 1933): 60–61.

25. *American Cotton Planter* (Montgomery), May 1854, 148; ibid., November 1854, 334; Leavitt, "Attempts to Improve Cattle Breeds," 63; *American Cotton Planter,* January 1854, 18–19; ibid., March 1855, 81; ibid., March 1858, 87.

26. *American Cotton Planter,* January 1854, 54; ibid., May 1854, 157.

27. Ibid., December 1856, 370, 376.

28. J. Crawford King, "The Closing of the Southern Range: An Exploratory Study," *Journal of Southern History* 48 (February 1982): 56.

29. U.S. Department of State, Office of the Census, *Sixth Census of the United States, 1840: Compendium, III* (Washington, D.C.: Government Printing Office, 1841); U.S. Department of the Interior, Office of the Census, *Eighth Census of the United States, 1860: Agriculture, II* (GPO, 1864).

30. Owsley, "Pattern of Migration," 149, 154.

31. *Sixth Census of the U.S.: Compendium; Eighth Census of the U.S.: Agriculture.*

32. Washington, Covington, and Baldwin counties, AMS Census 1850 and 1860.

33. Randolph, Washington, and Baldwin counties, AMS Census 1850 and 1860.

34. Covington and Washington counties, AMS Census 1850 and 1860.

35. Hilliard, *Hog Meat and Hoecake,* 4, 45, 124–30, 105.

36. Silas L. Loomis, "Distribution and Movement of Neat Cattle in the United States," in *Report of the Commissioner of Agriculture for the Year 1863* (GPO, 1864), 248–64; Hilliard, *Hog Meat and Hoecake,* 129.

37. Hilliard, *Hog Meat and Hoecake,* 121; *Report of the Commissioner of Patents for the Year 1855* (Washington, D.C., 1856), 405; *Mobile Daily Commercial Register and Patriot,* scattered issues of November and December 1859.

38. *Sixth Census of the U.S.: Compendium; Eighth Census of the U.S.: Agriculture;* Lowndes County, AMS Census 1850 and 1860.

39. *War of the Rebellion: A Compilation of the Official Records of the Union and Confederate Armies,* 70 vols. in 127 books and index (GPO, 1880–1901), series I, 16, pt. 1, 473.

40. Lauderdale and DeKalb counties, AMS Census 1860.

41. W. McLean to Mary, Holderfield Civil War Letters, Malcolm C. McMillan Microfilm Collection, Auburn University; Garrett to Gov. Thomas H. Watts, 4 June 1864, Governor Thomas Hill Watts Unprocessed Administrative Records, 1860–1865, Alabama Department of Archives and History, Montgomery; J. T. Trowbridge, *A Picture of the Desolated States and the Work of Restoration, 1865–1868* (Hartford, Connecticut: L. S. Stebbins, 1868), 435.

42. *Eighth Census of the U.S.: Agriculture;* U.S. Department of the Interior, Office of the Census, *Ninth Census of the United States, 1870: Wealth and Industry, III* (GPO, 1872).

43. John T. Milner, *Alabama: As It Was, As It Is, and As It Will Be* (Montgomery: Barrett and Brown, 1876); John T. Milner, quoted in Ethel Armes, *The Story of Coal and Iron in Alabama* (Birmingham: Chamber of Commerce, 1910), 269–70.

44. McWhiney, "Revolution in Nineteenth-Century Alabama Agriculture," 21; McDonald and McWhiney, "Antebellum Southern Herdsman," 163.

45. Frank L. Owsley, *Plain Folk of the Old South* (Baton Rouge: Louisiana State University Press, 1949); *Eighth Census of the U.S.: Agriculture*.

46. Loomis, "Distribution and Movement of Neat Cattle," 264, 258; Paul W. Gates, *Agriculture and the Civil War*, The Impact of the Civil War Series (New York: Alfred A. Knopf, 1965), 6; John Solomon Otto, *Southern Agriculture during the Civil War Era, 1860–1880*, Contributions in American History, no. 153 (Westport, Connecticut: Greenwood Press, 1994), 24.

47. Robert H. McKenzie, "The Economic Impact of Federal Operations in Alabama during the Civil War," *Alabama Historical Quarterly* 38 (Spring 1976): 55–56.

48. Walter L. Fleming, *Civil War and Reconstruction in Alabama* (New York: Columbia University Press, 1905), 75, 257; *OR* I, 10, 164–65; *OR* I, 16, pt. 1, 603.

49. *OR* I, 16, pt. 1, 605, 358, 495.

50. Ibid., 358, 390.

51. Ibid., 474–75.

52. *OR* I, 10, 292, 204, 212; *Huntsville Confederate*, 4 February 1863; *OR* I, 16, pt. 1, 350; *Montgomery Advertiser*, 18 December 1862.

53. *OR* I, 23, pt. 1, 249; *Selma Evening Reporter*, 27 May 1864.

54. *OR* I, 30, pt. 4, 547–48; *OR* I, 32, pt. 2, 645; *OR* I, 32, pt. 3, 718, 794, 718.

55. *OR* I, 23, pt. 2, 675; Governors' Letter Books, Alabama, May 1863–April 1865, Alabama Department of Archives and History, Montgomery, 21 April 1864, 7 July 1864.

56. McKenzie, "Economic Impact," 61.

57. *Biennial Report of the Comptroller of Public Accounts, Alabama* (Montgomery, 1861); *OR* I, 23, pt. 2, 840; *OR* I, 26, 163; *OR* I, 32, pt. 2, 552.

58. H. E. Owens to T. H. Watts, 22 June 1864, Governor Watts Unprocessed Administrative Records, 1860–1865, Alabama Department of Archives and History; Citizens of Pike County to T. H. Watts, 12 June 1864, ibid.

59. Allen W. Jones, "Unionism and Disaffection in South Alabama: The Case of Alfred Holley," *Alabama Review* 24 (April 1971): 127–28; Elez. P. Holley to T. H. Watts, 15 June 1864, Governor Watts Unprocessed Administrative Records, 1860–1865, Alabama Department of Archives and History.

60. *Acts of Alabama: 1864*, 165–66.

61. John H. Harper, "Rousseau's Alabama Raid," M.A. Thesis, Auburn University, 1965; McKenzie, "Economic Impact," 58.

62. C. Robert Watkins, "General Wilson's Raid through Alabama and Georgia in 1865," M.A. Thesis, Auburn University, 1959; Trowbridge, *Picture of the Desolated States*, 437.

63. *OR* I, 23, pt. 2, 816–17.

64. Thomas Robson Hay, "Lucius B. Northrop: Commissary General of the Confederacy," *Civil War History* 9 (March 1963): 12.

65. Ibid., 26; *OR* IV, 3, 249–50.

66. *OR* I, 24, pt. 3, 990–91.

NOTES TO PAGES 39–46

67. *OR* IV, 2, 470, 653, 836; *OR* IV, 3, 686.

68. *Southern Cultivator* (Athens, Georgia), November–December 1863, 133; *OR* IV, 3, 37; T. H. Watts to Jefferson Davis, 8 March 1864, Thomas H. Watts Collection, box 256, folder 22, Alabama Department of Archives and History; Watts to James A. Seddon, 12 April 1864, Governors' Letter Books, Alabama, May 1863–April 1865, Alabama Department of Archives and History.

69. Watts to Seddon, 12 December 1863, Governors' Letter Books, Alabama, May 1863–April 1865, Alabama Department of Archives and History; Watts to Seddon, 12 April 1864, ibid.; *Montgomery Advertiser,* 18 December 1862; *OR* IV, 3, 501–2; *Acts of Alabama: 1864,* 12; Malcolm C. McMillan, *The Disintegration of a Confederate State: Three Governors and Alabama's Wartime Home Front, 1861–1865* (Macon, Georgia: Mercer University Press, 1986), 86.

70. *Southern Cultivator,* November–December 1863, 133; *OR* IV, 3, 250, 1170.

71. *Montgomery Mail,* 15 July 1863; *OR* IV, 2, 575.

72. McMillan, *Disintegration of a Confederate State,* 190; *Southern Cultivator,* February 1865, 19. The Confederacy's original daily meat ration of three-quarters pound of pork or bacon or one and one-quarter pound of fresh or salt beef was often reduced to one-half that amount or less by war's end. Richard D. Goff, *Confederate Supply* (Durham: Duke University Press, 1969), 17.

73. "Report of the Statistician," in *Report of the Commissioner of Agriculture for the Year 1866* (GPO, 1867), 68; Loomis, "Distribution and Movement of Neat Cattle," 250–51. The 1870 census of agriculture is notoriously inaccurate, even more so than previous efforts. It is likely that the actual number of cattle exceeded that reported, but by how much is unknown. Information for some districts, for example, the cattle-dense St. Stephen's area of Washington County, was not reported.

74. The eight counties with over 10,000 cattle were Jackson, Choctaw, Clarke, Henry, Marengo, Monroe, Pike, and Wilcox. Five of seven counties with less than 5,000 head were northern counties: Colbert, Etowah, Marion, St. Clair, and Talladega. The other two were the central counties of Bibb and Lowndes. *Census of Agriculture, 1870*; Hilliard, *Hog Meat and Hoecake,* 131.

75. Otto, *Southern Agriculture,* 74, 100; *Eighth Census of the U.S.: Agriculture; Ninth Census of the U.S.: Wealth and Industry.*

76. McDonald and McWhiney, "The South from Self-Sufficiency to Peonage: An Interpretation," *American Historical Review* 85 (December 1980); McWhiney, "Revolution in Nineteenth-Century Alabama Agriculture," 3.

3. Agricultural Progressivism and the South

1. For McWhiney's and McDonald's arguments, see Forrest McDonald and Grady McWhiney, "The Antebellum Southern Herdsman: A Reinterpretation," *Journal of Southern History* 41 (May 1975): 147–66, and McWhiney, "The Revolution in Nineteenth-Century Alabama Agriculture," *Alabama Review* 31 (January 1978).

2. Statistics gathered from 1880 agricultural manuscript census schedules of Baldwin, Washington, and DeKalb counties.

3. Baldwin and Covington counties, AMS Census 1880.

4. DeKalb County, AMS Census 1880.

5. Winston County, AMS Census 1880.

6. Madison and Lauderdale counties, AMS Census 1880.
7. Marengo County, AMS Census 1880.
8. DeKalb, Baldwin, Marengo, and Lauderdale counties, AMS Census 1880.
9. J. Allen Tower, "Alabama's Shifting Cotton Belt," *Alabama Review* 1 (January 1948): 33–34; U.S. Department of the Interior, Census Office, *Tenth Census of the United States, 1880: Agriculture* (Washington, D.C.: Government Printing Office, 1883); U.S. Department of Commerce, Bureau of the Census, *Thirteenth Census of the United States, 1910: Agriculture* (GPO, 1914). See also Pete Daniel, *Breaking the Land: The Transformation of Cotton, Tobacco, and Rice Cultures since 1880* (Urbana: University of Illinois Press, 1985); Gilbert C. Fite, *Cotton Fields No More: Southern Agriculture, 1865–1980* (Lexington: University of Kentucky Press, 1984); Jack Temple Kirby, *Rural Worlds Lost: The American South, 1920–1960* (Baton Rouge: Louisiana State University Press, 1987).
10. *Tenth Census of the U.S.: Agriculture; Thirteenth Census of the U.S.: Agriculture.*
11. J. Crawford King, Jr., "The Closing of the Southern Range: An Exploratory Study," *Journal of Southern History* 48 (February 1982): 56.
12. Ibid., 56–58; *Acts of Alabama* (1866–1867), 61; *Acts of Alabama* (1878–1879), 298–301; *Acts of Alabama* (1890–1891), 1151–55.
13. King, "Closing of the Southern Range," 59; *Acts of Alabama* (1880–1881), 163–65.
14. King, "Closing of the Southern Range," 60; *Acts of Alabama* (1903 General), 431–38.
15. Steven Hahn, *The Roots of Southern Populism: Yeoman Farmers and the Transformation of the Georgia Upcountry, 1850–1890* (New York: Oxford University Press, 1983), 239–43.
16. William Warren Rogers, *The One-Gallused Rebellion: Agrarianism in Alabama, 1865–1896* (Baton Rouge: Louisiana State University Press, 1970); Sheldon Hackney, *Populism to Progressivism in Alabama* (Princeton, New Jersey: Princeton University Press, 1969). Neither Rogers nor Hackney mentions the possible influence of the open-range controversy. See also Theodore Saloutos, *Farmer Movements in the South, 1865–1933* (Berkeley: University of California Press, 1960).
17. *Opelika Times*, 5 September 1884, 26 September 1884, 19 September 1884. See also *Opelika Times*, 24 October 1884, and *Huntsville Gazette*, 17 January 1885.
18. *Opelika Times*, 24 October 1884; *Montgomery Advertiser*, 14 March 1901.
19. Jasper *Mountain Eagle*, 3 January 1912, 10 January 1912, 24 January 1912.
20. Ibid., 7 February 1912, 5 February 1913, 12 February 1913.
21. Hackney, *Populism to Progressivism*, 193. Hackney observes that the poll tax in effect disfranchised perhaps more than 20 percent of the white male population between the ages of twenty-one and forty-five. See also the chapter on the 1901 constitution in William Warren Rogers, Robert David Ward, Leah Rawls Atkins, and Wayne Flynt, *Alabama: The History of a Deep South State* (Tuscaloosa: University of Alabama Press, 1994).
22. *Clear Creek Lumber Co. v. Gossom*, 151 Ala. 450; *Ryall v. Allen*, 143 Ala. 222; *George et al. v. Chickasaw Land Co.*, 209 Ala. 648. See also *Dismukes v. Jones*, 151 Ala. 441.
23. *Joiner v. Winston*, 68 Ala. 129.
24. *Price v. DeKalb County*, 188 Ala. 419. See also *Dennis v. Chilton County*, 192 Ala. 146; *Hutto et al. v. Walker County*, 185 Ala. 505; *Edwards et al. v. Bibb County Board of Commissioners*, 193 Ala. 554; *Browning v. St. Clair County*, 195 Ala. 121; *State ex rel. Potts v. Court of County Commissioners of Lauderdale County*, 210 Ala. 508; *Campbell et al. v. Jefferson County*, 216 Ala. 251.

25. King, "Closing of the Southern Range," 62; *Mobile & Ohio Railroad Company v. Malone,* 46 Ala. 391; *Zeigler v. South & North Alabama Railroad Company,* 58 Ala. 594.

26. King, "Closing of the Southern Range," 62; *Mobile & Ohio Railroad Company v. Malone,* 46 Ala. 391; *Zeigler v. South & North Alabama Railroad Company,* 58 Ala. 594. Another common source of litigation was injury to livestock by neighbors or other people. The reports from these trials are among the most colorful involving livestock. In one Clay County case involving allegations of livestock and property destruction, the plaintiff's girlfriend accused the defendant of having "a hankering for her." In her estimation the defendant had planned to court her by keeping her boyfriend busy with repairing and replacing his destroyed property. *Burgess v. The State,* 44 Ala. 190. See also *Northcot v. The State,* 43 Ala. 330.

27. Benjamin D. Warfield to George W. Jones, "Livestock Correspondence, August–September 1918," Alabama State Council of Defense Program Administrative Files, 1917–1919, Alabama Department of Archives and History, Montgomery; N. B. Cravy to Council of Defense, "Livestock Correspondence, December 1918," ibid.

28. Norwood Allen Kerr, *A History of the Alabama Agricultural Experiment Station, 1883–1983* (Auburn: Auburn University, Alabama Agricultural Experiment Station, 1985), 3, 4.

29. Ibid., 4, 5.

30. Ibid., 5, 8.

31. Roy V. Scott, *The Reluctant Farmer: The Rise of Agricultural Extension to 1914* (Urbana: University of Illinois Press, 1970), 33; Kerr, *Alabama Agricultural Experiment Station,* 15, 18–19.

32. Kerr, *Alabama Agricultural Experiment Station,* 20–21; Bulletins 1, 5, 8, 9, and 15 of the Canebrake Agricultural Experiment Station, Canebrake Agricultural Experiment Station Records, Auburn University Archives. See also "Second Annual Report of the Canebrake Agricultural Experiment Station" (1889), Canebrake Agricultural Experiment Station Records, Auburn University Archives.

33. Kerr, *Alabama Agricultural Experiment Station,* 38; J. F. Duggar, *Feeding and Grazing Experiments with Beef Cattle,* Alabama Agricultural Experiment Station Bulletin No. 128 (Auburn: Alabama Polytechnic Institute, 1904), 54. After 1896, agricultural experimentation was also done at Tuskegee, a black land grant institution. See B. D. Mayberry, *The Role of Tuskegee University in the Origins, Growth and Development of the Negro Cooperative Extension System, 1881–1990* (Tuskegee: Tuskegee University Press, 1989).

34. Dan T. Gray and W. F. Ward, *Raising Beef Cattle in Alabama,* A.A.E.S. Bulletin No. 150 (API, 1910), 8; Gray and Ward, *Wintering Steers in Alabama,* A.A.E.S. Bulletin No. 151 (API, 1910); Gray and Ward, *Steer Feeding in Alabama,* A.A.E.S. Bulletin No. 163 (API, 1912), 61; Gray and Ward, *Raising and Fattening Beef Calves in Alabama,* A.A.E.S. Bulletin No. 177 (API, 1914), 71–72. The Aberdeen-Angus breed, more popularly called simply Angus, originated in northern Scotland. The solid black cattle began to be imported by North American cattlemen in the last quarter of the nineteenth century. M. E. Ensminger, *Beef Cattle Science* (Danville, Illinois: Interstate, 1968), 51–52.

35. C. A. Cary, *Texas or Tick Fever,* A.A.E.S. Bulletin No. 141 (API, 1907), 109, 111, 121, 156. See also Henry C. Dethloff and Donald H. Dyal, *A Special Kind of Doctor: A History of Veterinary Medicine in Texas* (College Station: Texas A & M University Press, 1991); Cecil Kirk Hutson, "Texas Fever in Kansas, 1866–1930," *Agricultural History* 68 (Winter 1994):

74–104; Elizabeth Diane Schafer, "Reveille for Professionalism: Alabama Veterinary Medical Association, 1907–1952," M.A. Thesis, Auburn University, 1988, 59.

36. *Montgomery Advertiser,* "Farm and Immigration Section," 25 November 1917, 2.

37. *Montgomery Advertiser,* 16 June 1911; *Acts of Alabama* (1907 General), 582–83.

38. This biographical information comes from three papers delivered at a centennial celebration of Cary's arrival at Auburn held in November 1992: Jack Simms, "Charles Allen Cary, the Citizen"; Wilford S. Bailey, "Charles Allen Cary: The Man and His Legacy"; and Calvin W. Schwabe, "Man of the New South—Charles Allen Cary: Auburn's Rural Health Pioneer." Copies in possession of author.

39. Schwabe, "Man of the New South," 4.

40. Simms, "Charles Allen Cary, the Citizen," 11; Bailey, "Charles Allen Cary: The Man and His Legacy," 8.

41. *Montgomery Advertiser,* 16 June 1911. The fifteen counties were Butler, Dallas, Elmore, Jackson, Lauderdale, Limestone, Lowndes, Madison, Marengo, Monroe, Montgomery, Perry, Pickens, Sumter, and Wilcox. Quoted headline also from above newspaper. *Montgomery Advertiser,* 8 January 1915, 1–2.

42. *Montgomery Advertiser,* 8 January 1915, 1–2; Schafer, "Reveille for Professionalism," 72.

43. *Montgomery Advertiser,* 4 April 1915, 6; ibid., 6 June 1915, 11; ibid., "Farm and Immigration Section," 12 December 1915, 1.

44. Cary, *Texas or Tick Fever,* 172; H. M. Graybill and W. M. Lewallen, *The Biology or Life History of the Cattle Tick as Determined at Auburn, Ala.,* and C. A. Cary, *Dipping Vats and Dips,* A.A.E.S. Bulletin No. 171 (API, 1913), 97–100.

45. F & I, 25 November 1917, 7; *Acts of Alabama* (1919 General), 29–32.

46. See John Whiteclay Chambers II, *The Tyranny of Change: America in the Progressive Era, 1890–1920,* St. Martin's Series in 20th Century U.S. History (New York: St. Martin's Press, 1992); Robert H. Wiebe, *The Search for Order, 1877–1920,* The Making of America Series (New York: Hill and Wang, 1967); William A. Link, *The Paradox of Southern Progressivism, 1880–1930,* Fred W. Morrison Series in Southern Studies (Chapel Hill: University of North Carolina Press, 1992); Edward L. Ayers, *The Promise of the New South: Life after Reconstruction* (New York: Oxford University Press, 1992).

47. Link, *Paradox of Southern Progressivism,* xi–xii; T. A. McEwen, interview with author, Hanover, Alabama, 2 October 1995; James Cravy, interview with author, Florala, Alabama, 19 October 1995.

48. *Fort Payne Journal,* 6 August 1913; Link, *Paradox of Southern Progressivism,* xi–xii.

49. *Dothan Weekly Eagle,* 11 July 1919; *Gadsden Daily Times-News,* 8 July 1919.

50. *Cullman Democrat,* 17 April 1919 and 24 April 1919; *Gadsden Daily Times-News,* 5 July 1919 and 8 August 1919.

51. *Hagan et al. v. State ex rel. Batchelor,* 207 Ala. 514; *Cary, State Veterinarian v. Commissioners Court of Clarke County,* 218 Ala. 23.

52. *Progressive Farmer* (Georgia and Alabama edition), 23 November 1929, 9.

53. Daniel, *Breaking the Land,* 7 (quote), 6–8.

54. Kerr, *Alabama Agricultural Experiment Station,* 39–40; Daniel, *Breaking the Land,* 9. The Farmers' Institute, inaugurated at Auburn in 1889 and at Tuskegee in 1892, comprised a series of meetings held around the state in which lectures and demonstrations were given by experiment station experts. Usually attracting about 100 visitors, the institutes operated

until World War I. W. E. Hinds, *Boll Weevil in Alabama,* A.A.E.S. Bulletin No. 188 (API, 1916), 23. Responding to Alabama farmers' interest in the cattle industry, Alabama Polytechnic Institute first offered a sophomore-level course on judging beef cattle during the 1912–1913 academic year, familiarizing future extension agents with the intricacies of these increasingly popular animals. The Animal Husbandry Department added a course in beef cattle production shortly after World War I. *Catalogue of the Alabama Polytechnic Institute, 1912–1913* (Auburn, 1913), 159; *Catalogue of the Alabama Polytechnic Institute, 1919–1920* (Auburn, 1920).

55. Kerr, *Alabama Agricultural Experiment Station,* 40–41; Hinds, *Boll Weevil in Alabama,* plate 1; Daniel, *Breaking the Land,* 16 (quote).

56. Kathryn Holland Braund, " 'Hog Wild' and 'Nuts': Billy Boll Weevil Comes to the Alabama Wiregrass," *Agricultural History* 63 (Summer 1989): 22–25; *Andalusia Star,* 11 June 1915 and 25 April 1916.

57. Hazel Stickney, "The Conversion from Cotton to Cattle Economy in the Alabama Black Belt, 1930–1960," Ph.D. Diss., Clark University, Worcester, Massachusetts, 1961, 9–12. Only ten counties can claim substantial Black Belt acreage: Dallas, Lowndes, Sumter, Marengo, Montgomery, Greene, Perry, Hale, Bullock, and Wilcox. For an explanation of Black Belt soils, see George D. Scarseth, *Morphological, Greenhouse, and Chemical Studies of the Black Belt Soils of Alabama,* A.A.E.S. Bulletin No. 237 (API, 1932). See also J. Sullivan Gibson, "The Alabama Black Belt: Its Geographic Status," *Economic Geography* 17 (January 1941): 1–23; Merle Prunty, Jr., "Recent Quantitative Changes in the Cotton Regions of the Southeastern States," *Economic Geography* 27 (July 1951): 189–208.

58. D. G. Sturkie and Clarence M. Wilson, *Alfalfa Production in Alabama,* A.A.E.S. Bulletin No. 300 (API, 1956), 3.

59. Stickney, "Conversion from Cotton to Cattle," 55.

60. F & I, 25 March 1917; *Progressive Farmer,* January 1965, 25.

61. Thomas McAdory Owen, *History of Alabama and Dictionary of Alabama Biography,* vol. 2 (Chicago: S. J. Clark, 1921), 1096; *American Aberdeen-Angus Herd Book,* vol. 37 (Webster City, Iowa: American Aberdeen-Angus Breeders' Association, 1930), v.

62. Owen, *History of Alabama,* vol. 1, 59–60, 677; F & I, 14 November 1915, 10; John M. Hazelton, *History and Hand Book of Hereford Cattle and Hereford Bull Index* (Kansas City: Hereford Journal Company, 1925), 174; *Alabama Official and Statistical Register, 1915* (Montgomery: Alabama Department of Archives and History, 1915), 139. As the name suggests, the Hereford breed originated in the English county of Hereford. The Hereford, which was first imported to the United States in the early nineteenth century, is a red animal with a white face and underline. Thus these cattle are often referred to as "white faced" cattle. Ensminger, *Beef Cattle Science,* 62.

63. Owen, *History of Alabama,* vol. 1, 516–17; Hazelton, *History and Hand Book of Hereford Cattle,* 173; F & I, 25 March 1917, 6.

64. *Alabama Blue Book and Social Register, 1929* (Birmingham: Blue Book Publishing Company, 1929), 379, 333, 327; "The Proceedings of the Twentieth Annual Meeting of the Alabama Live Stock Association, February 10–11, 1916," file 33, Livestock Pamphlets, Lists, and Forms, Alabama State Council of Defense: Program Administrative Files, 1917–1919, Alabama Department of Archives and History; *American Aberdeen-Angus Herd Book,* vol. 37, v.

65. *Alabama Blue Book, 1929,* 322; F & I, 12 December 1915, 1; *Progressive Farmer,* 11 January 1930, 5.

66. Glenn N. Sisk, "Alabama's Black Belt: A Social History, 1875–1917," Ph.D. Diss., Duke University, 1951, 471; *American Short-Horn Herd Book,* vol. 89 (Chicago: American Shorthorn Breeders' Association, 1916), xxvii; *Montgomery Advertiser,* 19 October 1910, 12; F & I, 10 October 1915, 1; F & I, 6 March 1917, 6; F & I, 14 November 1915, 13. Shorthorns, native cattle of northeastern England, come in a variety of colors, including roan, red, and white. The first beef breed brought to the United States, Shorthorns were first imported shortly after the American Revolution. Ensminger, *Beef Cattle Science,* 68–70. Though Kentucky is not generally considered a midwestern state, its stock-raising practices, especially those of the Bluegrass region, resembled the system then in place in the Midwest.

67. *Greensboro Watchman,* 2 May 1918; Stickney, "Conversion from Cotton to Cattle," 176, 191.

68. Rudolf Alexander Clemen, *The American Livestock and Meat Industry* (New York: Ronald Press, 1923), 440; Hinds, *Boll Weevil in Alabama,* 23–24. Since the greatest part of the weevil's destruction occurred in the hot summer months, experiments found that planting early or sowing early-maturing strains helped prevent massive losses. Tower, "Alabama's Shifting Cotton Belt," 35; Gibson, "Alabama Black Belt," 20.

69. Stickney, "Conversion from Cotton to Cattle," 42; *Montgomery Advertiser,* 20 October 1910; *Progressive Farmer,* 7 May 1921, 6.

70. *Thirteenth Census of the U.S.: Agriculture;* Census Bureau, *Fourteenth Census of the United States, 1920: Agriculture,* vol. 6, pt. 2 (GPO, 1922).

71. Sisk, "Alabama's Black Belt," 470; Stickney, "Conversion from Cotton to Cattle," 74–76.

72. *Montgomery Advertiser,* 20 October 1910; F & I, 10 October 1915, 5; F & I, 11 October 1915, 6.

73. *Montgomery Advertiser,* 19 October 1910, 12. The Marion Junction livestock show, promoted by M. F. Smith, S. A. Brice, W. V. Waugh, and others, awarded beef breed ribbons to A. B. Moore, A.D. Simmons, H. P. Randall, and R. S. Tubbs. *Montgomery Advertiser,* 20 October 1910, 2.

74. *Montgomery Advertiser,* 24 November 1915, 1, 5; F & I, 14 November 1915, 10; F & I, 25 November 1916, 9. County fairs began to feature purebred cattle shows and sales around this time as well. The 1915 Limestone County Fair offered a sale of fifty Shorthorn and Angus cattle owned by Cobb and Derby. F &I, 7 October 1915.

75. Purebred, or thoroughbred, livestock do not have to be registered; registered cattle, however, must be thoroughbred. The registering process generally involves pedigree documents and a fee paid to the breed association. Registered stock often receive much better prices than purebred but unregistered stock. Leonard William Brinkman, Jr., "The Historical Geography of Improved Cattle in the United States to 1870," Ph.D. Diss., University of Wisconsin, 1964, vol. 2, 84, 90, 126, 138; U.S. Department of the Interior, Office of the Census, *Eleventh Census of the United States, 1890: Report of the Statistics of Agriculture in the United States* (GPO, 1895).

76. *American Hereford Record and Hereford Herd Book,* vol. 23 (Columbia, Missouri: American Hereford Cattle Breeders' Association, 1901), 436; *American Aberdeen-Angus Herd-Book,* vol. 13 (Davenport, Iowa: American Aberdeen-Angus Breeders' Association, 1904),

442; *American Aberdeen-Angus Herd-Book,* vol. 16 (1907), xxxiii; *American Short-Horn Herd Book,* vol. 70 (1908), x–xxv, vol. 89 (1916), xiv–lxvi. Though this issue listed thirty-two Alabama cattlemen, I was unable to ascertain the place of residence for two of them. Black Belt cattlemen constituted fifteen of the remaining thirty. *American Hereford Record and Hereford Herd Book,* vol. 43 (1917), v; *American Aberdeen-Angus Herd Book,* vol. 27 (Chicago, 1917), 1.

77. *American Hereford Record and Hereford Herd Book,* vol. 43 (1917), v; *American Short-Horn Herd Book,* vol. 89 (1916), xiv–lxvi.

78. F & I, 12 December 1915, 8; F & I, 17 October 1915, 23.

4. The Midwestern Model Meets the South

1. J. F. Duggar, "Annual Report of Farm Demonstration Work, 1917," box 117, Alabama Cooperative Extension Service Records, Auburn University Archives; *Breeder's Gazette* (Chicago), 3 October 1918, 534–35. The Red Poll is a hornless, red English breed of cattle.

2. "Livestock Wanted, October 10, 1918" and "Livestock for Sale, October 10, 1918," in "Livestock Correspondence, October 1–15, 1918," Alabama State Council of Defense Program Administrative Files, 1917–1919, Alabama Department of Archives and History, Montgomery.

3. Pete Daniel, *Breaking the Land: The Transformation of Cotton, Tobacco, and Rice Cultures since 1880* (Urbana: University of Illinois Press, 1985), 16.

4. *Montgomery Advertiser,* "Farm and Immigration" section, 11 March 1917, 2; F & I, 11 November 1917, 3.

5. F & I, 28 January 1917, 1; *Demopolis Times,* 4 April 1918; F & I, 25 November 1917, 2.

6. M. J. Danner, *Livestock Marketing Agencies in Alabama,* Alabama Agricultural Experiment Station Bulletin No. 284 (Auburn: Alabama Polytechnic Institute, 1952), 6.

7. F & I, 14 October 1917, 11; F & I, 9 December 1917, 2; F & I, 11 November 1917, 6; Marianna Snow to Mrs. E. H. Wilson, 19 March 1975, letter in possession of the Alabama Cattlemen's Association, Montgomery.

8. *Montgomery Advertiser,* 6 June 1918, 1.

9. Snow to Wilson, ACA, Montgomery; "Annual Report, 1929," file "Animal Industry, 1929," box 136, ACES.

10. U.S. Department of Commerce, Bureau of the Census, *Thirteenth Census of the United States, 1910: Agriculture* (Washington, D.C.: Government Printing Office, 1914); Census Bureau, *Fourteenth Census of the United States, 1920: Agriculture,* vol. 6, pt. 2 (GPO, 1922).

11. *Alabama Jersey News* (Auburn), 1944, 7; "Ledger, Accounts: 1910–1913," file 9, box 1, Bermuda Polled Hereford Farm Records, Auburn University Archives.

12. *Progressive Farmer,* 9 April 1921, 21; F & I, 23 December 1917, 3; John Parrish and William Hunt Eaton, "History of the Alabama Dairy Industry and Development, 1918–1967," unpublished paper, W. H. Eaton Collection, Auburn University Archives.

13. William Hunt Eaton, "Alabama Jersey Cattle Club," W. H. Eaton Collection; *Alabama Jersey News,* 1949, 29.

14. Stewart E. Tolnay and E. M. Beck, "Rethinking the Role of Racial Violence in the Great Migration," in Alferdteen Harrison, ed., *Black Exodus: The Great Migration from the American South* (Jackson: University Press of Mississippi, 1991), 22.

15. Harry Simms, State Agent for Negro Men, "Annual Report of Extension Work in Alabama, 1923," box 117, ACES.

16. Ibid.

17. T. M. Campbell, "Second Annual Report of the Agricultural Extension Service as Performed by Negroes for the Year Ending December 31, 1917," box 117, ACES.

18. James R. Grossman, "Black Labor Is the Best Labor: Southern White Reactions to the Great Migration," in Harrison, *Black Exodus*, 55; *Southern Plantation* (Montgomery), 3 February 1875, 51. *Southern Plantation* was the official organ of the state Grange in Alabama.

19. P. O. Davis, "The Negro Exodus and Southern Agriculture," *American Review of Reviews* 68 (October 1923): 402–6; Grossman, "Black Labor," 55.

20. Hazel Stickney, "The Conversion from Cotton to Cattle Economy in the Alabama Black Belt, 1930–1960," Ph.D. Diss., Clark University, Worcester, Massachusetts, 1961, 2; Davis, "Negro Exodus," 405.

21. Price histories supplied to the author by the Alabama Agricultural Statistics Service, Montgomery; *Thirteenth Census of the U.S.: Agriculture; Fourteenth Census of the U.S.: Agriculture*.

22. *Breeder's Gazette,* 19 December 1918, 1115.

23. Alabama Agricultural Statistics Service.

24. Census Bureau, *U.S. Census of Agriculture: 1935,* vol. 1, pt. 2, Southern States (GPO, 1936); Stickney, "Conversion from Cotton to Cattle," 7.

25. J. C. Grimes, *Steer Feeding Experiments in the Black Belt of Alabama,* A.A.E.S. Bulletin No. 231 (API, 1930), 4; "Animal Industry, 1924," box 108, ACES; "Animal Industry, 1926," box 125, ACES.

26. "Animal Industry, 1928," box 131, ACES; "Animal Industry, 1929," box 136, ACES.

27. T. Whit Athey, Jr., interview with Meg Crawford, 20 August 1993; Dr. M. L. Crawford, interview with Meg Crawford, 18 August 1993. Both interviews on videotape and in the possession of the ACA, Montgomery.

28. Merle Prunty, Jr., "Recent Quantitative Changes in the Cotton Regions of the Southeastern States," *Economic Geography* 27 (July 1951): 194.

29. Norwood Allen Kerr, *A History of the Alabama Agricultural Experiment Station, 1883–1983* (Auburn: Auburn University, Alabama Agricultural Experiment Station, 1985), 56.

30. *Progressive Farmer,* 9 November 1929, 16; *Highlights of Agricultural Research* (Alabama Agricultural Experiment Station, Auburn) 5 (Winter 1958): 8.

31. Stickney, "Conversion from Cotton to Cattle," 158.

32. J. D. Moore to Mr. Smith, 4 October 1932, "Cattle," box 93, ACES; personal information obtained from Alabama Extension Service Personnel Files, Duncan Hall, Auburn University. Moore was also a cattle raiser; he and his brother-in-law owned a Lowndes County farm twenty miles southwest of Montgomery. Moore to Mr. Smith, 24 October 1934, "Livestock," box 94, ACES.

33. "Animal Husbandry, 1931," box 142, ACES; "Animal Husbandry, 1932," box 149, ACES.

34. "Animal Husbandry, 1932," box 149, ACES; "A History of Livestock in Alabama," anonymous manuscript in "Rural Alabama: Pre and Post World War II" file, folder 27, box 2, ACES.

35. Moore to Leroy Upshaw, 19 November 1932, "Cattle," box 93, ACES; Moore to Judge Stonewall McConnico, 29 November 1932, ibid.; Joe Patton to E. M. Cothran, 4 August 1933, ibid.

36. Moore to Dr. Walter Sorrell, 29 August 1932, ibid.; Moore to S. C. Hill, 18 April 1932, ibid.; Moore to W. R. Odom and R. B. Donovan, 3 December 1934, "Livestock," box 94, ACES; Moore to Fred Pickett, 6 February 1933, "Cattle," box 93, ACES; Moore to O. Harris and Son, 28 January 1933, "Cattle," box 93, ACES; V. W. Lewis to Moore, 26 December 1933, "Cattle," box 93, ACES; Moore to S. Willis Van Meter, 13 August 1932, "Cattle," box 93, ACES.

37. G. B. Phillips to Moore, 22 November 1932, "Cattle," box 93, ACES; Moore to Phillips, 25 November 1932, ibid.; Moore to J. T. Belue, 11 September 1933, ibid.; Moore to W. C. Vail, 24 May 1935, ibid.; Moore to E. E. Hale, 17 April 1933, ibid.; Moore to Hale, 28 May 1932, ibid.; Moore to C. L. Hollingsworth, 16 May 1932, ibid.

38. Moore to R. J. Goode, 29 November 1932, ibid.; G. S. Evans to U. C. Jenkins, 6 October 1934, "Livestock," box 94, ACES; Madison Jones to Moore, 18 December 1933, "Cattle," box 93, ACES; Moore to Overton Harris, 13 December 1933, "Cattle," box 93, ACES. Other information acquired from numerous letterheads in this collection.

39. Moore to R. B. Burgess, 13 September 1933, "Livestock Producers Assoc.: 1934," box 93, ACES; Moore to Directors of Livestock Producers Association, 5 May 1934, ibid.; List of Officers and Directors of the Livestock Producers Association, March 1933, ibid.

40. Daniel, *Breaking the Land*, 92–93; *U.S. Census of Agriculture*, Alabama, 1935; Census Bureau, *U.S. Census of Agriculture: 1945*, vol. 1, Statistics for Counties, pt. 21, Alabama (GPO, 1946). Farmers often took their poorest land out of production, resulting in a less dramatic decrease in cotton production. For instance, while total acreage declined by 40 percent between 1929 and 1934, production declined by only 29 percent.

41. Stickney, "Conversion from Cotton to Cattle," 75; Moore to Maxcel Voorhies, 15 December 1932, "Livestock Producers Assoc.: 1934," box 93, ACES; *Progressive Farmer*, January 1933, 8.

42. *Birmingham News*, 14 July 1936; Gilbert C. Fite, "Southern Agriculture since the Civil War: An Overview," *Agricultural History* 53 (January 1979): 16; *Montgomery Advertiser*, 2 January 1944, 3; David E. Hamilton, *From New Day to New Deal: American Farm Policy from Hoover to Roosevelt, 1928–1933* (Chapel Hill: University of North Carolina Press, 1991), 232; *Progressive Farmer*, March 1934, 16.

43. Hamilton, "New Day to New Deal," 232; *Montgomery Advertiser*, 2 January 1944, 3; Athey, Jr., interview with Meg Crawford.

44. C. Roger Lambert, "The Drought Cattle Purchase, 1934–1935: Problems and Complaints," *Agricultural History* 45 (April 1971): 85; Theodore Saloutos, *The American Farmer and the New Deal* (Ames: Iowa State University Press, 1982), 193–94.

45. Lambert, "Drought Cattle Purchase," 89; Saloutos, *American Farmer*, 194; Moore to Bledsoe and Vail, 28 August 1934, "Livestock Producers Assoc.: 1934," box 93, ACES; Moore to Henry A. Wallace, 24 September 1934, "Livestock," box 94, ACES; R. C. McChord to Moore, 29 September 1934, "Livestock Producers Assoc.: 1934," box 93, ACES.

46. Moore to F. C. Clapp, 21 September 1934, "Livestock," box 94, ACES; *Birmingham News*, 1 August 1936.

47. "A History of Livestock in Alabama," box 2, ACES; "Animal Husbandry, 1935," box 157, ACES; *Handbook of Alabama Agriculture*, 6th ed. (Auburn: Alabama Extension Service, 1958), 262.

48. Kerr, *Alabama Agricultural Experiment Station*, 62; Robert L. Geiger, Jr., *A Chronological History of the Soil Conservation Service and Related Events* (Washington, D.C.: United States Department of Agriculture, 1955), 3; *Progressive Farmer*, 15–30 April 1931, 33.

49. Kerr, *Alabama Agricultural Experiment Station*, 62; "Annual Report: Animal Husbandry, 1942," box 203, ACES; *Birmingham News*, 5 June 1937, 26 June 1937, 3 April 1937.

50. Geiger, *Chronological History of the Soil Conservation Service*, 1–12; *Birmingham News*, 6 June 1936.

51. Saloutos, *American Farmer*, 237–38.

52. *Progressive Farmer*, December 1941, 10.

53. *Birmingham News*, 19 August 1941, 11 October 1941, 7 March 1942.

54. Moore to Directors of the Livestock Producers Association, "Livestock Producers Assoc.: 1934," box 93, ACES; *Birmingham News*, 19 February 1938, 12 November 1938.

55. *Statutes at Large of the United States of America*, vol. 49, pt. 1 (GPO, 1936), 436–39; William H. Johnson, Jr., interview with Meg Crawford, c. 1993; Athey, Jr., interview with Meg Crawford.

56. Christiana McFayden Campbell, *The Farm Bureau and the New Deal: A Study of the Making of National Farm Policy, 1933–1940* (Urbana: University of Illinois Press, 1962), 58, 88, 102; George B. Tindall, *The Emergence of the New South, 1913–1945* (Baton Rouge: Louisiana State University Press, 1967), 131, 392, 427; *Progressive Farmer*, November 1933, 8.

57. Fite, "Southern Agriculture since the Civil War," 20–21.

58. "Animal Husbandry, 1935," box 157, ACES; "A History of Livestock in Alabama," box 2, ACES.

59. "A History of Livestock in Alabama," box 2, ACES; *Birmingham News*, 19 September 1936, 13 May 1937, 14 November 1936; Wendell H. Stepp and Pamela Ann Stepp, *Dothan: A Pictorial History* (Norfolk, Virginia: Donning, 1984), 160. The fourth livestock show of 1937 was held at Hartselle in Morgan County.

60. "A History of Livestock in Alabama," box 2, ACES; "Animal Husbandry, 1938," box 171, ACES; *Birmingham News*, 15 October 1938; "Annual Report of Extension Animal Husbandry," file "Animal Industry: Livestock, Dairying, 1940," box 188, ACES.

61. "A History of Livestock in Alabama," box 2, ACES; William H. Gregory, "History of Alabama's Beef Cattle Industry," Program for 11th Annual Meeting of the ACA, 12–13 February 1954, ACA, Montgomery.

62. *Livestock Breeder Journal*, 1 February 1970, 28; ibid., July 1982, 50; *Birmingham News*, 15 October 1938, 8 October 1938, 6 June 1936.

63. William H. McDonald, *Wadsworth Flats: The Story of Three Remarkable Brothers* (Prattville, Alabama: Twin Valley Press, 1983), 17, 19, 21, 52.

64. *Selma Times Journal*, 29 June 1941.

65. Ibid.

66. Ibid.; "Lee Place Accounts," Ledger of C. A. Webb, Jr., in possession of the ACA, Montgomery.

67. *Birmingham News*, 4 February 1939, 1 April 1939, 28 August 1937; *Alabama Official*

and Statistical Register, 1967 (Montgomery: Alabama Department of Archives and History, 1967), 322.

68. Mack Maples, interview with Meg Crawford, 22 January 1994; "Limestone Personalities," *Limestone Legacy* 2 (July 1980): 100–101, Limestone County Historical Society, Athens, Alabama.

69. *U.S. Census of Agriculture,* Alabama, 1935 and 1945; W. K. McPherson, *A General Appraisal of the Livestock Industry in the Southeastern States,* A.A.E.S. Bulletin No. 257 (API, 1942), 4, 6, 13.

70. J. D. Moore to N. Moore, 18 April 1934, "Railroad: Correspondence," box 93, ACES; Ben F. Alvord, M. A. Crosby, and E. G. Schiffman, *Factors Influencing Alabama Agriculture,* A.A.E.S. Bulletin No. 250 (API, 1941), 60; *U.S. Census of Agriculture,* Alabama, 1935 and 1945. The ten core Black Belt counties are Bullock, Dallas, Greene, Hale, Lowndes, Marengo, Montgomery, Perry, Sumter, and Wilcox. The Black Belt average of ten head of cattle per farm is deceptive in that the large number of sharecroppers owning a single milk cow brings down the average significantly. Therefore comparing the herds of those who actually owned cattle in the Black Belt with the herds of those in the hills would most likely reflect an even wider discrepancy.

71. *U.S. Census of Agriculture,* Alabama, 1935 and 1945; "Animal Industry: Livestock, Dairying, 1940," box 188, ACES.

72. "Animal Industry: Livestock, Dairying, 1940," box 188, ACES; *U.S. Census of Agriculture,* Alabama, 1935 and 1945.

73. Alvord et al., *Factors Influencing Alabama Agriculture,* 60; *Farm Production and Marketing in Alabama,* Alabama Extension Service Circular No. 241 (API, 1943), 18; James Cravy, interview with the author, 19 October 1995.

74. *Acts of Alabama* (1939), 187–96.

5. Cattle in the Cotton Fields

1. Gilbert C. Fite, *Cotton Fields No More: Southern Agriculture, 1865–1980* (Lexington: University Press of Kentucky, 1984), 197.

2. U.S. Department of Commerce, Bureau of the Census, *U.S. Census of Agriculture, 1974,* vol. 1, pt. 1, Alabama, State and County Data (Washington, D.C.: Government Printing Office, 1977); Census Bureau, *U.S. Census of Agriculture, 1992,* vol. 1, pt. 1, Alabama, State and County Data (GPO, 1994).

3. Gilbert C. Fite, "Southern Agriculture since the Civil War: An Overview," *Agricultural History* 53 (January 1979): 18–19.

4. "Annual Report, 1943, Extension Animal Husbandry," box 209, Alabama Cooperative Extension Service Records, Auburn University Archives.

5. Ibid.; "A History of Livestock in Alabama," anonymous manuscript in "Rural Alabama: Pre and Post World War II" file, folder 27, box 2, ACES.

6. Information obtained from various undated clippings from the *Birmingham Post* and other papers in "Annual Report, 1943, Extension Animal Husbandry"; *Birmingham Post,* 20 May 1944, 3.

7. Newspaper clippings in "Annual Report, 1943, Extension Animal Husbandry"; "Annual Report, 1944, Animal Husbandry," box 215, ACES.

8. "Annual Report, 1943, Extension Animal Husbandry"; *Birmingham Post,* 8 April 1944, 3; *Birmingham News,* 4 May 1942, 3.

9. Prices supplied by the Alabama Agricultural Statistics Service, Montgomery.

10. Ibid.; Census Bureau, *U.S. Census of Agriculture, 1945,* vol. 1, Statistics for Counties, pt. 21, Alabama (GPO, 1946).

11. *U.S. Census of Agriculture, 1945,* Alabama.

12. *Birmingham News,* 15 January 1944, 5; Dan W. Hollis, *It's Great to Be Number One: The Dynamic Story of the Alabama Cattlemen's Association* (Montgomery: Alabama Cattlemen's Association, 1985), 6; *Birmingham Post* article in "Annual Report, 1943, Extension Animal Husbandry."

13. U. C. Jenkins to Mr. Cattleman, 29 December 1943, form letter in possession of the Alabama Cattlemen's Association, Montgomery; Hollis, *Great to Be Number One,* 6–7.

14. Hollis, *Great to Be Number One,* 7.

15. *Alabama Official and Statistical Register, 1943* (Wetumpka: Alabama Department of Archives and History, 1943), 258.

16. Ibid., 317.

17. Hollis, *Great to Be Number One,* 8; William H. Johnson, Jr., interview with Meg Crawford, c. 1993.

18. Alabama Cooperative Extension Service Biographies (ACES BIOS), Auburn University Archives.

19. Hollis, *Great to Be Number One,* 8.

20. *Progressive Farmer,* 28 August 1926, 8; Norwood Allen Kerr, *A History of the Alabama Agricultural Experiment Station, 1883–1983* (Auburn: Auburn University, Alabama Agricultural Experiment Station, 1985), 56.

21. Census Bureau, *Fourteenth Census of the United States, 1920: Agriculture,* vol. 6, pt. 2 (GPO, 1922); *Farm Production and Marketing in Alabama,* Alabama Extension Service Circular 241 (Auburn: Alabama Polytechnic Institute, 1943), 8–9; *Progressive Farmer,* January 1942, 14.

22. *Fourteenth Census of the U.S.: Agriculture.*

23. "Annual Dairy Report (County Agents), 1950," box 195, ACES; Riley G. Arnold Papers, box 93, ACES; Kerr, *History of the Alabama Agricultural Experiment Station,* 73.

24. "Annual Dairy Report (County Agents), 1950"; Riley G. Arnold Papers.

25. Wayne Flynt, *Mine, Mill, and Microchip: A Chronicle of Alabama Enterprise* (Northridge, California: Windsor Publications, 1987), 294.

26. Ibid., 294–95.

27. Lemuel Morrison, "John Lemly Morrison: President, Dairy Fresh Corporation," J. Lem Morrison Collection, Auburn University Archives.

28. Ibid.; "Minutes of Organization Meeting of the American Dairy Association of Alabama," file 3, box 1, American Dairy Association of Alabama Records: 1950–1982, Auburn University Archives.

29. Census Bureau, *U.S. Census of Agriculture, 1964,* vol. 1, State and County Data, pt. 32, Alabama (GPO, 1967); Census Bureau, *U.S. Census of Agriculture,* Alabama, 1945.

30. Census Bureau, *U.S. Census of Agriculture,* Alabama, 1945.

31. Fite, *Cotton Fields No More,* 183.

32. "Dr. R. S. Sugg," box 95, ACES; "Annual Report of W. H. Gregory," in "Annual

Report: Marketing—1949," box 244, ACES; T. Whit Athey, Jr., interview with Meg Crawford, 20 August 1993; *Alabama Cattleman*, May 1968, 7.

33. "Annual Report of W. H. Gregory"; "Annual Narrative Report," in "Marketing 1953, Mr. Gregory–Mr. Farquhar," box 266, ACES; *Sylacauga Advertiser*, 21 July 1949; *Florence Times*, 8 February 1949.

34. "Annual Narrative Report" (1953); Gregory to County Agents, 7 May 1953, in "Marketing 1953, Mr. Gregory–Mr. Farquhar"; "Minutes of Joint Meeting of the Committee Arrangement with Mr. P. O. Davis," Alabama Cattlemen's Association Records: 1944–1958 (microfilm, Auburn University Archives).

35. Cattle and price estimates supplied by the Alabama Agricultural Statistics Service; "Cattle Prices," in "Marketing 1953, Mr. Gregory–Mr. Farquhar."

36. "Annual Narrative Report, 1958," box 295, ACES; Alabama Agricultural Statistics Service.

37. *Highlights of Agricultural Research* (Alabama Agricultural Experiment Station, Auburn) (Winter 1958): 8; ibid. (Summer 1957): 4; Fite, *Cotton Fields No More*, 197.

38. Fite, *Cotton Fields No More*, 195; L. E. Ensminger and E. M. Evans, *Establishment and Maintenance of White Clover-Grass Pastures*, A.A.E.S. Bulletin No. 327 (API, 1960), 3.

39. *Highlights of Agricultural Research* (Spring 1962): 6; Fite, *Cotton Fields No More*, 197.

40. Hazel Stickney, "The Conversion from Cotton to Cattle Economy in the Alabama Black Belt, 1930–1960," Ph.D. Diss., Clark University, Worcester, Massachusetts, 1961, 126; M. J. Danner, *Livestock Marketing Agencies in Alabama*, A.A.E.S. Bulletin No. 284 (API, 1952), 4.

41. M. J. Danner, *How Alabama Farmers Buy and Sell Livestock*, A.A.E.S. Bulletin No. 281 (API, 1952), 13–15; Danner, *Livestock Marketing Agencies in Alabama*, 7. Country buyers and local dealers were especially popular in poor, isolated farming areas, because these marketers generally drove their trucks from farm to farm to pick up cattle. One negative result of this practice was the tendency of the country buyer to pay a fixed amount per head—in the absence of scales—which more often than not cost the farmer money. Since the local dealer markets were geared toward hog production, these were found almost exclusively in the Wiregrass region by the 1950s.

42. E. Ham Wilson, interview with Dwayne Cox, 23 July 1992.

43. Ibid.

44. Hollis, *Great to Be Number One*, 44–48. The ACA has held annual elections since its founding. E. Ham Wilson, interview with Dwayne Cox, 5 October 1992.

45. *Acts of Alabama* (Special Regular Session 1951), vol. 2, 1295–96, 1496–99; Wilson interview, 5 October 1992.

46. *Acts of Alabama* (1951), vol. 1, 409–15; Hollis, *Great to Be Number One*, 121–22.

47. *Acts of Alabama* (1951), 266; Johnson, interview with Meg Crawford; *U.S. Census of Agriculture*, Alabama, 1945 and 1964.

48. Starting in 1949 all ACA letterheads were embossed with the longhorn head. "Alabama Cattlemen's Association Records: 1944–1958," reel 1, ACA Records (microfilm, Auburn University Archives).

49. Hollis, *Great to Be Number One*, 120; Wilson interview, 5 October 1992.

50. Hollis, *Great to Be Number One*, 10–11, 65; Wilson interview, 23 July 1992.

51. Hollis, *Great to Be Number One*, 276–79.

52. Hollis, *Great to Be Number One*, 280; Wilson interview, 23 July 1992.

53. Hollis, *Great to Be Number One*, 280; Wilson interview, 23 July 1992 and 11 November 1992; *Alabama Cattleman*, June 1962, 12. The referendum vote was 156,214 for and 82,648 against, with only Chambers, Randolph, and Wilcox counties voting against. Close votes were also recorded in several hill counties such as Walker, Winston, and Cleburne.

54. Hollis, *Great to Be Number One*, 44–48. The ACA's twelfth president, Mortimer Jordan IV, though a resident of Tuscaloosa and Birmingham, owned a Black Belt cattle ranch, as did Tuscaloosa resident O. J. Henley, the group's fifth president. In addition, both leaders from southwestern Alabama also owned Black Belt ranches.

55. *U.S. Census of Agriculture*, Alabama, 1964.

56. Stickney, "Conversion from Cotton to Cattle," 96–99.

57. Ibid., 109–20.

58. Ibid., 170–75; Merle Prunty, Jr., "The Renaissance of the Southern Plantation," *Geographical Review* 45 (October 1955): 459–91; Charles S. Aiken, "The Fragmented Neoplantation: A New Type of Farm Operation in the Southeast," *Southeastern Geographer* 11 (April 1971): 43–51.

59. Stickney, "Conversion from Cotton to Cattle," 187–88.

60. William H. McDonald, *Wadsworth Flats: The Story of Three Remarkable Brothers* (Prattville: Twin Valley Press, 1983), 17, 21, 52; "Annual Narrative Report, Animal Husbandry, 1958," box 295, ACES.

61. Stickney, "Conversion from Cotton to Cattle," 103–6.

62. Ibid., 176–82, 191–96; *Alabama Cattleman*, November 1960, 5.

63. *Alabama Cattleman*, January 1965, 5.

64. Ibid., January 1969, 8 and January 1967, 5.

65. Hollis, *Great to Be Number One*, 46, 48.

66. Stickney, "Conversion from Cotton to Cattle," 24; P. O. Davis, "The Negro Exodus and Southern Agriculture," *American Review of Reviews* 68 (October 1923): 401–7.

6. New Farmers in the New South

1. *Birmingham News*, 22 January 1938; M. J. Danner, *Livestock Marketing Agencies in Alabama*, Alabama Agricultural Experiment Station Bulletin No. 284 (Auburn: Alabama Polytechnic Institute, 1952), 6, 9; Jerry M. Martin and M. J. Danner, *The Alabama Slaughter Cattle Industry*, A.A.E.S. Bulletin No. 365 (Auburn: Auburn University, 1966), 4, 11.

2. J. R. Meadows and M. J. Danner, *Movement of Cattle and Calves through Alabama Auction Markets*, A.A.E.S. Bulletin No. 360 (Auburn University, 1965), 3, 6, 7, 11; *Alabama Cattleman*, December 1962, 22.

3. "Annual Narrative Report, 1958," box 295, Alabama Cooperative Extension Service Records, Auburn University Archives; *Farm Journal* (Philadelphia), April 1959, 96.

4. *Farm Journal*, April 1959, 37, 96; *Alabama Cattleman*, February 1959, 7; *Alabama Cattleman*, December 1962, 22.

5. *Alabama Cattleman*, October 1966, 70; ibid., July 1966, 8; ibid., March 1962, 12; ibid., January 1967, 9; E. Ham Wilson, interview with Dwayne Cox, 23 July 1992.

6. *Alabama Cattleman*, April 1968, 10; ibid., May 1968, 7; ibid., December 1960, 7; ibid., March 1963, 7; "Annual Report for Agricultural Production, Management and Natural Resources Use," in "Annual Narrative Report, 1964," box 311, ACES; Dan W. Hollis, *It's*

NOTES TO PAGES 148–156

Great to Be Number One: The Dynamic Story of the Alabama Cattlemen's Association (Montgomery: Alabama Cattlemen's Association, 1985), 185–88.

7. Wilson interview, 23 July 1992; *Alabama Cattleman*, March 1963, 7.

8. Joe A. Akerman, Jr., *American Brahman: A History of the American Brahman* (Houston: American Brahman Breeders Association, 1982), 231; *Birmingham Post*, 23 December 1944; Gilbert C. Fite, *Cotton Fields No More: Southern Agriculture, 1865–1980* (Lexington: University Press of Kentucky, 1984), 197. Generally gray or red with drooping ears, a long face, and a hump over the shoulders, the Brahman is an amalgamation of several Indian types with a strain of European breeding. M. E. Ensminger, *Beef Cattle Science* (Danville, Illinois: Interstate, 1968, 4th ed.), 102–3.

9. Akerman, *American Brahman*, 329.

10. Hollis, *Great to Be Number One*, 109; Extension Radio Show, 16 April 1958, transcript in "Alabama Extension Service Program in Beef Cattle," in "Annual Narrative Report, 1958," box 295, ACES; Ensminger, *Beef Cattle Science*, 99, 102–3.

11. *Alabama Cattleman*, October 1960, 9; Ladd Haystead and Gilbert C. Fite, *The Agricultural Regions of the United States* (Norman: University of Oklahoma Press, 1955), 128.

12. Hollis, *Great to Be Number One*, 110–11; *Alabama Cattleman*, September 1967, 21.

13. *Alabama Cattleman*, June 1975, 27; ibid., July 1975, 23.

14. *Alabama Cattleman*, October 1964, 5; Hollis, *Great to Be Number One*, 22; *Alabama Cattleman*, September 1964, 3.

15. Hollis, *Great to Be Number One*, 28–30; *Alabama Cattleman*, September 1975, 11; *Alabama Cattleman*, July 1972, 9.

16. Hollis, *Great to Be Number One*, 44–57; "1960 ACA Membership Report," reel 2 (1959–1970), ACA Records (microfilm, Auburn University Archives); "1975 ACA Final Membership List," reel 3 (1971–1980), ACA Records.

17. U.S. Department of Commerce, Bureau of the Census, *U.S. Census of Agriculture, 1964*, vol. 1, State and County Data, pt. 32, Alabama (Washington, D.C.: Government Printing Office, 1967); Census Bureau, *U.S. Census of Agriculture, 1974*, vol. 1, pt. 1, Alabama, State and County Data (GPO, 1977).

18. *Alabama Cattleman*, October 1960, 6.

19. Fite, *Cotton Fields No More*, 188.

20. *Alabama Cattleman*, October 1960, 22; ibid., July 1961, 19; T. A. McEwen, interview with the author, 2 October 1995.

21. *Cullman Tribune*, 25 January 1968; *Sand Mountain Reporter* (Albertville-Boaz), 30 January 1964; *Sand Mountain Reporter*, 6 February 1964.

22. Bessie Odom, interview with Jenni Wallace, 24 February 1990, interview no. 61, American Folklore and Oral History Records, Auburn University Archives; Pauline Blevins, interview with the author, 24 October 1995.

23. Hollis, *Great to Be Number One*, 106–7; "Complete List and Sale Days of Alabama's Livestock Markets, 1970," reel 2 (1959–1970), ACA Records; "1980 Beef Promotion Stockyard Collection Report," reel 3 (1971–1980), ACA Records.

24. *U.S. Census of Agriculture*, Alabama, 1964 and 1974.

25. Census Bureau, *U.S. Census of Agriculture, 1982*, vol. 1, pt. 1, Alabama, State and County Data (GPO, 1984); Census Bureau, *U.S. Census of Agriculture, 1992*, vol. 1, pt. 1, Alabama, State and County Data (GPO, 1994).

26. *U.S. Census of Agriculture,* Alabama, 1964.
27. *U.S. Census of Agriculture,* Alabama, 1992.
28. Ibid.
29. E. Ham Wilson, interview with Dwayne Cox, 11 November 1992; U. C. Jenkins to Mr. Cattleman, 29 December 1943, in possession of the ACA, Montgomery; "Minutes of Meeting of Board of Directors," reel 1 (1944–1958), ACA Records.
30. Alabama Agricultural Statistics Service, Montgomery.
31. *Farm Implement News,* 29 January 1948, 60, 36–37; Brooks Robert Blevins, "Fallow Are the Hills: A Century of Rural Modernization in the Arkansas Ozarks," M.A. Thesis, Auburn University, 1994, 43.
32. *Progressive Farmer,* March 1953, 119; *Implement & Tractor,* 31 January 1963, B-94; *U.S. Census of Agriculture,* Alabama, 1964; *Farm Implement News,* 10 May 1950, 50; *Progressive Farmer,* May 1962, 39.
33. *Farm Technology,* April 1964, 35; *Progressive Farmer,* May 1970, 46–47.
34. *Farm Journal,* January 1974, 23; ibid., January 1975, 24–25.
35. Between 1987 and 1996 five of ten ACA presidents owned Black Belt ranches.
36. *Alabama Cattleman,* July 1990, 24, 27; Wilson interview, 11 November 1992; *Alabama Cattleman,* May 1991, 5; Jeremy Rifkin, *Beyond Beef: The Rise and Fall of the Cattle Culture* (New York: Penguin, 1992).
37. *Alabama Cattleman,* April 1990, 46; ibid., December 1990, 15; Dale L. Huffman and W. Russell Egbert, *Advances in Lean Ground Beef Production,* A.A.E.S. Bulletin No. 606 (Auburn: Auburn University, 1990).
38. Hollis, *Great to Be Number One,* 56; Alabama Agricultural Statistics Service; *Alabama Cattleman,* April 1994, 14; *Alabama Cattleman,* December 1985, 15.

BIBLIOGRAPHY

Primary Sources

Newspapers and Magazines

Alabama Cattleman
Alabama Jersey News
American Cotton Planter
Andalusia Star
Birmingham News
Birmingham Post
Breeder's Gazette (Chicago)
Cullman Democrat
Cullman Tribune
Demopolis Times
Dothan Weekly Eagle
Farm Implement News
Farm Journal (Philadelphia)
Farm Technology
Florence Times
Fort Payne Journal
Gadsden Daily Times-News
Greensboro Watchman
Highlights of Agricultural Research (Auburn)
Huntsville Confederate
Huntsville Gazette
Implement & Tractor
Mobile Daily Commercial Register and Patriot
Montgomery Advertiser
Montgomery Mail
Mountain Eagle (Jasper)
Opelika Times
Progressive Farmer
Sand Mountain Reporter
Selma Evening Reporter
Selma Times Journal
Southern Cultivator
Southern Plantation
Sylacauga Advertiser

Other Primary Sources

Acts of Alabama
Agricultural Manuscript Census Schedules: Alabama. Selected Counties, 1850, 1860, and 1880.
Alabama Blue Book and Social Register, 1929. Birmingham: Blue Book Publishing Company, 1929.
Alabama Cattlemen's Association Records. Microfilm, Auburn University Archives.
Alabama Cooperative Extension Service Records. Auburn University Archives.
Alabama Official and Statistical Register, 1915. Montgomery: Alabama Department of Archives and History, 1915.

BIBLIOGRAPHY

Alabama Official and Statistical Register, 1943. Wetumpka: Alabama Department of Archives and History, 1943.

Alabama Official and Statistical Register, 1967. Montgomery: Alabama Department of Archives and History, 1967.

Alabama State Council of Defense Program Administrative Files. Alabama Department of Archives and History, Montgomery.

Alvord, Ben F., M. A. Crosby, and E. G. Schiffman. *Factors Influencing Alabama Agriculture.* Alabama Agricultural Experiment Station (A.A.E.S.) Bulletin No. 250. 1941.

American Aberdeen-Angus Herd Book. Numerous volumes. Webster City, Iowa: American Aberdeen-Angus Breeders' Association.

American Dairy Association of Alabama Records. Auburn University Archives.

American Short-Horn Herd Book. Numerous volumes. Chicago: American Shorthorn Breeders' Association.

Athey, T. Whit, Jr. Interview with Meg Crawford. 20 August 1993. Alabama Cattlemen's Association, Montgomery.

Bartram, William. *Travels through North & South Carolina, Georgia, East & West Florida, the Cherokee Country, the Extensive Territories of the Muscogulges, or Creek Confederacy, and the Country of the Chactaws.* Philadelphia: James & Johnson, 1791. Reprinted as *William Bartram Travels* with an introduction by Robert McCracken Peck. Salt Lake City: Peregrine Smith, 1980.

Bermuda Polled Hereford Farm Records. Auburn University Archives.

Biennial Report of the Comptroller of Public Accounts, Alabama. Montgomery, 1861.

Blevins, Pauline. Personal interview. 24 October 1995.

Browning v. St. Clair County, 195 Ala. 121. *Alabama Reports (AR).*

Burgess v. The State, 44 Ala. 190. *AR.*

Campbell et al. v. Jefferson County, 216 Ala. 251. *AR.*

Canebrake Agricultural Experiment Station Records. Auburn University Archives.

Cary, C. A. *Dipping Vats and Dips.* A.A.E.S. Bulletin No. 171. 1913.

———. *Texas or Tick Fever.* A.A.E.S. Bulletin No. 141. 1907.

Cary, State Veterinarian v. Commissioners Court of Clarke County, 218 Ala. 23. *AR.*

Catalogue of the Alabama Polytechnic Institute, 1912–1913. Auburn, 1913.

Catalogue of the Alabama Polytechnic Institute, 1919–1920. Auburn, 1920.

Claiborne, J. F. H. "A Trip through the Piney Woods." *Publications of the Mississippi Historical Society* 9 (1906): 487–538.

Clay, Clement Claiborne. Papers. Duke University Special Collections, Durham, North Carolina.

Clear Creek Lumber Co. v. Gossom, 151 Ala. 450. *AR.*

Cravy, James. Personal interview. 19 October 1995.

Crawford, Dr. M. L. Interview with Meg Crawford. 18 August 1993. Alabama Cattlemen's Association, Montgomery.

Danner, M. J. *How Alabama Farmers Buy and Sell Livestock.* A.A.E.S. Bulletin No. 281. 1952.

———. *Livestock Marketing Agencies in Alabama.* A.A.E.S. Bulletin No. 284. 1952.

Dennis v. Chilton County, 192 Ala. 146. *AR.*

Dent, James Horry. Farm Journals and Account Books. Microfilm and photocopies in Auburn University Archives. Originals in University of Alabama Special Collections.

Dismukes v. Jones, 151 Ala. 441. *AR.*

BIBLIOGRAPHY

Duggar, J. F. *Feeding and Grazing Experiments with Beef Cattle.* A.A.E.S. Bulletin No. 128. 1904.
Eaton, William Hunt. William Hunt Eaton Collection. Auburn University Archives.
Edwards et al. v. Bibb County Board of Commissioners, 193 Ala. 554. *AR.*
Ensminger, L. E., and E. M. Evans. *Establishment and Maintenance of White Clover-Grass Pastures.* A.A.E.S. Bulletin No. 327. 1960.
Farm Production and Marketing in Alabama. Alabama Extension Service Circular No. 241. Auburn: Alabama Polytechnic Institute, 1943.
George et al. v. Chickasaw Land Co., 209 Ala. 648. *AR.*
Gosse, Philip Henry. *Letters from Alabama.* London: Morgan and Chase, 1859.
Governors' Letter Books, Alabama, May 1863–April 1865. Alabama Department of Archives and History, Montgomery.
Gray, Dan T., and W. F. Ward. *Raising and Fattening Beef Calves in Alabama.* A.A.E.S. Bulletin No. 177. 1914.
———. *Raising Beef Cattle in Alabama.* A.A.E.S. Bulletin No. 150. 1910.
———. *Steer Feeding in Alabama.* A.A.E.S. Bulletin No. 163. 1912.
———. *Wintering Steers in Alabama.* A.A.E.S. Bulletin No. 151. 1910.
Graybill, H. M., and W. M. Lewallen. *The Biology or Life History of the Cattle Tick as Determined at Auburn, Ala.* A.A.E.S. Bulletin No. 171. 1913.
Grimes, J. C. *Steer Feeding Experiments in the Black Belt of Alabama.* A.A.E.S. Bulletin No. 231. 1930.
Hagan et al. v. State ex rel. Batchelor, 207 Ala. 514. *AR.*
Hamilton, William B. Papers. Duke University Special Collections, Durham, North Carolina.
Handbook of Alabama Agriculture. 6th ed. Auburn: Alabama Extension Service, 1958.
Hawkins, Benjamin. *A Combination of a Sketch of the Creek Country in the Years 1798 and 1799 by Benjamin Hawkins and Letters of Benjamin Hawkins, 1796–1806,* vol. 3, pt. 1 of *Collections of the Georgia Historical Society.* Savannah: Georgia Historical Society, 1848; Spartanburg, South Carolina: The Reprint Company, 1982.
———. Papers. Photocopies, Auburn University Archives.
Hinds, W. E. *Boll Weevil in Alabama.* A.A.E.S. Bulletin No. 188. 1916.
History and Hand Book of Hereford Cattle and Hereford Bull Index. Numerous volumes. Kansas City, Missouri: Hereford Journal Company.
Hodgson, Adam. *Letters from North America Written during a Tour in the United States and Canada,* vol. 1. London: Hurst, Robinson, & Co., 1824.
Holderfield Civil War Letters. Malcolm C. McMillan Microfilm Collection. Auburn University Library.
Huffman, Dale L., and W. Russell Egbert. *Advances in Lean Ground Beef Production.* A.A.E.S. Bulletin No. 606. 1990.
Hutto et al. v. Walker County, 185 Ala. 505. *AR.*
Johnson, William H., Jr. Interview with Meg Crawford. c. 1993. Alabama Cattlemen's Association, Montgomery.
Joiner v. Winston, 68 Ala. 129. *AR.*
McEwen, T. A. Personal interview. 2 October 1995.
McPherson, W. K. *A General Appraisal of the Livestock Industry in the Southeastern States.* A.A.E.S. Bulletin No. 257. 1942.

BIBLIOGRAPHY

Maples, Malcom "Mack." Interview with Meg Crawford. 22 January 1994. Alabama Cattlemen's Association, Montgomery.

Martin, Jerry M., and M. J. Danner. *The Alabama Slaughter Cattle Industry.* A.A.E.S. Bulletin No. 365. 1966.

Meadows, J. R., and M. J. Danner. *Movement of Cattle and Calves through Alabama Auction Markets.* A.A.E.S. Bulletin No. 360. 1965.

Melish, John. *Travels through the United States of America in the Years 1806 & 1807, and 1809, 1810, & 1811.* London: George Cowie and Co., 1818; New York: Johnson Reprint Corp., 1970.

Milner, John T. *Alabama: As It Was, As It Is, and As It Will Be.* Montgomery: Barrett and Brown, 1876.

Mobile & Ohio Railroad Company v. Malone, 46 Ala. 391. AR.

Morrison, John Lemly. J. Lem Morrison Collection. Auburn University Archives.

Northcot v. The State, 43 Ala. 330. AR.

Odom, Bessie. Interview with Jenni Wallace. 24 February 1990. Interview No. 61, American Folklore and Oral History Records, Auburn University Archives.

Panton, Leslie & Co. Papers. Microfilm, Auburn University Library. Original collection at the University of West Florida, Pensacola.

Park, John. Papers. Alabama Department of Archives and History, Montgomery.

Penicaut, M. *Annals of Louisiana from the Establishment of the First Colony under M. D'Iberville, to the Departure of the Author to France, in 1722*, trans. by B. F. French. Historical Collections of Louisiana and Florida. New York: J. Sabin & Sons, 1869.

Pope, John. *A Tour through the Southern and Western Territories of the United States of North America.* 1792. Reprinted with an introduction and indexes by J. Barton Starr. Bicentennial Floridiana Facsimile Series. Gainesville: University of Florida Press, 1979.

Price v. DeKalb County, 188 Ala. 419. AR.

Priestley, Herbert Ingram, ed. and trans. "The Luna Papers: Documents Relating to the Expedition of Don Tristan De Luna Y Arellano for the Conquest of La Florida in 1559–1661." *Publications of the Florida State Historical Society* 1, no. 8 (1928).

Report of the Commissioner of Agriculture for the Year 1863. Washington, D.C.: Government Printing Office (GPO), 1864.

Report of the Commissioner of Agriculture for the Year 1866. GPO, 1867.

Report of the Commissioner of Patents for the Year 1848. GPO, 1849.

Report of the Commissioner of Patents for the Year 1855. GPO, 1856.

Romans, Bernard. *A Concise Natural History of East and West Florida.* New York, 1775; New Orleans: Pelican Publishing, 1961.

Rowland, Dunbar, and A. G. Sanders, comps., eds., and trans. *French Dominion, 1704–1743*, vol. 3 of *Mississippi Provincial Archives.* Jackson: Press of the Mississippi Department of Archives and History, 1932.

Rowland, Mrs. Dunbar. "Peter Chester, Third Governor of the Province of West Florida under British Dominion, 1770–1781." *Publications of the Mississippi Historical Society.* 5 (1925): 1–183.

Ryall v. Allen, 143 Ala. 222. AR.

Scarseth, George D. *Morphological, Greenhouse, and Chemical Studies of the Black Belt Soils of Alabama.* A.A.E.S. Bulletin No. 237. 1932.

State ex rel. Potts v. Court of County Commissioners of Lauderdale County, 210 Ala. 508. AR.
Statutes at Large of the United States of America, GPO.
Stuart, James. *Three Years in North America*. Edinburgh: Robert Cadell, 1833.
Sturkie, D. G., and Clarence M. Wilson. *Alfalfa Production in Alabama*. A.A.E.S. Bulletin No. 300. 1956.
Tait Collection. Auburn University Archives.
Toulmin, Harry. "A Geographical and Statistical Sketch of the District of Mobile." *American Register* 6 (1810): 332–39.
Trowbridge, J. T. *A Picture of the Desolated States and the Work of Restoration, 1865–1868*. Hartford, Connecticut: L. S. Stebbins, 1868.
U.S. Department of Commerce, Bureau of the Census. *Thirteenth Census of the United States, 1910: Agriculture*. GPO, 1914.
———. *Fourteenth Census of the United States, 1920: Agriculture*. GPO, 1922.
———. *U.S. Census of Agriculture: 1935*, vol. 1, pt. 2, Southern States. GPO, 1936.
———. *U.S. Census of Agriculture: 1945*, vol. 1, pt. 21, Statistics for Counties, Alabama. GPO, 1946.
———. *U.S. Census of Agriculture, 1964*, vol. 1, pt. 32, State and County Data, Alabama. GPO, 1967.
———. *U.S. Census of Agriculture, 1974*, vol. 1, pt. 1, Alabama, State and County Data. GPO, 1977.
———. *U.S. Census of Agriculture, 1982*, vol. 1, pt. 1, Alabama, State and County Data. GPO, 1984.
———. *U.S. Census of Agriculture, 1992*, vol. 1, pt. 1, Alabama, State and County Data. GPO, 1994.
U.S. Department of the Interior. Office of the Census. *Eighth Census of the United States, 1860: Agriculture, II*. GPO, 1864.
———. *Ninth Census of the United States, 1870: Wealth and Industry, III*. GPO, 1872.
———. *Tenth Census of the United States, 1880: Agriculture*. GPO, 1883.
———. *Eleventh Census of the United States, 1890: Agriculture*. GPO, 1895.
U.S. Department of State. Office of the Census. *Sixth Census of the United States, 1840: Compendium, III*. GPO, 1841.
War of the Rebellion: A Compilation of the Official Records of the Union and Confederate Armies. 70 vols. in 127 books and index. GPO, 1880–1901.
Watts, Governor Thomas Hill. Unprocessed Administrative Records, 1860–1865. Alabama Department of Archives and History, Montgomery.
———. Thomas H. Watts Collection. Alabama Department of Archives and History, Montgomery.
Wilson, Edward Hamilton "Ham." Interviews with Dwayne Cox. 23 July 1992, 5 October 1992, 11 November 1992. Auburn University Archives.
Zeigler v. South & North Alabama Railroad Company, 58 Ala. 594. AR.

Secondary Sources

Aiken, Charles S. "The Fragmented Neoplantation: A New Type of Farm Operation in the Southeast." *Southeastern Geographer* 11 (April 1971): 43–51.
Akerman, Joe A., Jr. *American Brahman: A History of the American Brahman*. Houston: American Brahman Breeders Association, 1982.

BIBLIOGRAPHY

———. *Florida Cowman: A History of Florida Cattle Raising.* Kissimmee: Florida Cattlemen's Association, 1976.
Armes, Ethel. *The Story of Coal and Iron in Alabama.* Birmingham: Chamber of Commerce, 1910.
Arnade, Charles W. "Cattle Raising in Spanish Florida, 1513–1763." *Agricultural History* 35 (July 1961): 116–24.
Ayers, Edward L. *The Promise of the New South: Life after Reconstruction.* New York: Oxford University Press, 1992.
Bailey, Wilford S. "Charles Allen Cary: The Man and His Legacy." Unpublished paper presented at the centennial celebration of Cary's arrival at Auburn, November 1992. Copies in possession of the author.
Ball, T. H. *A Glance into the Great South-East, or, Clarke County, Alabama, and Its Surroundings, from 1540 to 1877.* Chicago: Knight & Leonard, 1879.
Blevins, Brooks Robert. "Fallow Are the Hills: A Century of Rural Modernization in the Arkansas Ozarks." M.A. Thesis, Auburn University, 1994.
Bonner, James C. "Profile of a Late Ante-Bellum Community." *American Historical Review* 49 (July 1944): 663–80.
Braund, Kathryn Holland. " 'Hog Wild' and 'Nuts': Billy Boll Weevil Comes to the Alabama Wiregrass." *Agricultural History* 63 (Summer 1989): 15–39.
Brewer, George E. "History of Coosa County." *Alabama Historical Quarterly* 4 (Spring 1942): 7–151.
Brinkman, Leonard William, Jr. "The Historical Geography of Improved Cattle in the United States to 1870." Ph.D. Diss., University of Wisconsin, Madison, 1964.
Campbell, Christiana McFayden. *The Farm Bureau and the New Deal: A Study of the Making of National Farm Policy, 1933–1940.* Urbana: University of Illinois Press, 1962.
Chambers, John Whiteclay II. *The Tyranny of Change: America in the Progressive Era, 1890–1920.* St. Martin's Series in 20th Century U.S. History. New York: St. Martin's Press, 1992.
Cherry, F. L. "The History of Opelika and Her Agricultural Tributary Territory." *Alabama Historical Quarterly* 15, nos. 2, 3, and 4 (1954).
Clark, Thomas D., and John D. W. Guice. *Frontiers in Conflict: The Old Southwest, 1795–1830.* Histories of the American Frontier. Albuquerque: University of New Mexico Press, 1989.
Clemen, Rudolf Alexander. *The American Livestock and Meat Industry.* New York: Ronald Press, 1923.
Cochran, John Perry. "James Asbury Tait and His Plantations." M.A. Thesis, University of Alabama, 1951.
Daniel, Pete. *Breaking the Land: The Transformation of Cotton, Tobacco, and Rice Cultures since 1880.* Urbana: University of Illinois Press, 1985.
Davis, P. O. "The Negro Exodus and Southern Agriculture." *American Review of Reviews* 68 (October 1923): 401–7.
Dethloff, Henry C., and Donald H. Dyal. *A Special Kind of Doctor: A History of Veterinary Medicine in Texas.* College Station: Texas A & M University Press, 1991.
Doran, Michael F. "Antebellum Cattle Herding in the Indian Territory." *Geographical Review* 66 (January 1976): 48–58.
Ensminger, M. E. *Beef Cattle Science.* Danville, Illinois: Interstate, 1968.

Fabel, Robin F. A. *The Economy of British West Florida, 1763–1783.* Tuscaloosa: University of Alabama Press, 1988.
Fite, Gilbert C. *Cotton Fields No More: Southern Agriculture, 1865–1980.* Lexington: University Press of Kentucky, 1984.
———. "Southern Agriculture since the Civil War: An Overview." *Agricultural History* 53 (January 1979): 3–21.
Fleming, Walter L. *Civil War and Reconstruction in Alabama.* 1905. Reprint. Gloucester, Massachusetts: Peter Smith, 1949.
Flynt, Wayne. *Mine, Mill, and Microchip: A Chronicle of Alabama Enterprise.* Northridge, California: Windsor Publications, 1987.
Fogel, Robert William, and Stanley L. Engerman. *Time on the Cross: The Economics of American Negro Slavery.* Boston: Little, Brown & Co., 1974.
Gates, Paul W. *Agriculture and the Civil War.* The Impact of the Civil War Series. New York: Alfred A. Knopf, 1965.
Geiger, Robert L., Jr. *A Chronological History of the Soil Conservation Service and Related Events.* Washington, D.C.: United States Department of Agriculture, 1955.
Genovese, Eugene D. "Livestock in the Slave Economy of the Old South: A Revised View." *Agricultural History* 36 (July 1962): 143–49.
Gibson, J. Sullivan. "The Alabama Black Belt: Its Geographic Status." *Economic Geography* 17 (January 1941): 1–23.
Goff, Richard D. *Confederate Supply.* Durham, North Carolina: Duke University Press, 1969.
Gray, Lewis Cecil. *History of Agriculture in the Southern United States to 1860.* 2 vols. Gloucester, Massachusetts: Carnegie Institution of Washington, 1958.
Guice, John D. W. "Cattle Raisers of the Old Southwest: A Reinterpretation." *Western Historical Quarterly* 8 (April 1977): 166–87.
Hackney, Sheldon. *Populism to Progressivism in Alabama.* Princeton, New Jersey: Princeton University Press, 1969.
Hahn, Steven. *The Roots of Southern Populism: Yeoman Farmers and the Transformation of the Georgia Upcountry, 1850–1890.* New York: Oxford University Press, 1983.
Hamilton, David E. *From New Day to New Deal: American Farm Policy from Hoover to Roosevelt, 1928–1933.* Chapel Hill: University of North Carolina Press, 1991.
Hamilton, Peter J. *Colonial Mobile.* 1897. Reprint. Tuscaloosa: University of Alabama Press, 1976.
———. *The Founding of Mobile, 1702–1718.* Mobile: Commercial Printing Co., 1911.
Harper, John H. "Rousseau's Alabama Raid." M.A. Thesis, Auburn University, 1965.
Harrison, Alferdteen, ed. *Black Exodus: The Great Migration from the American South.* Jackson: University Press of Mississippi, 1991.
Hay, Thomas Robson. "Lucius B. Northrop: Commissary General of the Confederacy." *Civil War History* 9 (March 1963): 5–23.
Haystead, Ladd, and Gilbert C. Fite. *The Agricultural Regions of the United States.* Norman: University of Oklahoma Press, 1955.
Higginbotham, Jay. *Old Mobile: Fort Louis de la Louisiane, 1702–1711.* Mobile: Museum of the City of Mobile, 1977.
Hilliard, Sam Bowers. *Hog Meat and Hoe Cake: Food Supply in the Old South, 1840–1860.* Carbondale: Southern Illinois University Press, 1972.

Hollis, Dan W. *It's Great to Be Number One: The Dynamic Story of the Alabama Cattlemen's Association.* Montgomery: Alabama Cattlemen's Association, 1985.

Hulbert, Archer B. *The Paths of Inland Commerce: A Chronicle of Trail, Road, and Waterway.* The Chronicles of America Series. New Haven: Yale University Press, 1920.

Hutson, Cecil Kirk. "Texas Fever in Kansas, 1866–1930." *Agricultural History* 68 (Winter 1994): 74–104.

Jones, Allen W. "Unionism and Disaffection in South Alabama: The Case of Alfred Holley." *Alabama Review* 24 (April 1971): 114–32.

Jordan, Terry G. *North American Cattle-Ranching Frontiers: Origins, Diffusion, and Differentiation.* Histories of the American Frontier. Albuquerque: University of New Mexico Press, 1993.

———. *Trails to Texas: Southern Roots of Western Cattle Ranching.* Lincoln: University of Nebraska Press, 1981.

Jordan, Weymouth T. *Hugh Davis and His Alabama Plantation.* University, Alabama: University of Alabama Press, 1948.

Kerr, Norwood Allen. *A History of the Alabama Agricultural Experiment Station, 1883–1983.* Auburn: Auburn University, Alabama Agricultural Experiment Station, 1985.

King, J. Crawford. "The Closing of the Southern Range: An Exploratory Study." *Journal of Southern History* 48 (February 1982): 53–70.

Kirby, Jack Temple. *Rural Worlds Lost: The American South, 1920–1960.* Baton Rouge: Louisiana State University Press, 1987.

Knox, John. *A History of Morgan County, Alabama.* Decatur: Morgan County Board of Revenue and Control, 1967.

Lackey, Richard S., and John D. W. Guice. *Frontier Claims in the Lower South.* New Orleans: Polyanthos, 1977.

Lambert, C. Roger. "The Drought Cattle Purchase, 1934–1935: Problems and Complaints." *Agricultural History* 45 (April 1971): 85–93.

Leavitt, Charles T. "Attempts to Improve Cattle Breeds in the United States, 1790–1860." *Agricultural History* 7 (April 1933): 51–67.

"Limestone Personalities." *Limestone Legacy* 2 (July 1980): 100–101. Athens, Alabama: Limestone County Historical Society.

Link, William A. *The Paradox of Southern Progressivism, 1880–1930.* Fred W. Morrison Series in Southern Studies. Chapel Hill: University of North Carolina Press, 1992.

McDonald, Forrest, and Grady McWhiney. "The Antebellum Southern Herdsman: A Reinterpretation." *Journal of Southern History* 41 (May 1975): 147–66.

———. "The South from Self-Sufficiency to Peonage: An Interpretation." *American Historical Review* 85 (December 1980): 1095–118.

McDonald, William H. *Wadsworth Flats: The Story of Three Remarkable Brothers.* Prattville, Alabama: Twin Valley Press, 1983.

McKenzie, Robert H. "The Economic Impact of Federal Operations in Alabama during the Civil War." *Alabama Historical Quarterly* 38 (Spring 1976): 51–68.

McMillan, Malcolm C. *The Disintegration of a Confederate State: Three Governors and Alabama's Wartime Home Front, 1861–1865.* Macon, Georgia: Mercer University Press, 1986.

McWhiney, Grady. "The Revolution in Nineteenth-Century Alabama Agriculture." *Alabama Review* 31 (January 1978): 3–32.

Mayberry, B. D. *The Role of Tuskegee University in the Origins, Growth and Development of the Negro Cooperative Extension System, 1881–1990.* Tuskegee: Tuskegee University Press, 1989.

Otto, John Solomon. "The Migration of the Southern Plain Folk: An Interdisciplinary Synthesis." *Journal of Southern History* 51 (May 1985): 183–200.

———. *Southern Agriculture during the Civil War Era, 1860–1880.* Contributions in American History, no. 153. Westport, Connecticut: Greenwood Press, 1994.

———. *The Southern Frontiers, 1607–1860: The Agricultural Evolution of the Colonial and Antebellum South.* Contributions in American History, no. 133. New York: Greenwood Press, 1989.

Owen, Thomas McAdory. *History of Alabama and Dictionary of Alabama Biography.* Chicago: S. J. Clark, 1921.

Owsley, Frank L. "The Pattern of Migration and Settlement on the Southern Frontier." *Journal of Southern History* 11 (May 1945): 147–76.

———. *Plain Folk of the Old South.* Baton Rouge: Louisiana State University Press, 1949.

Posey, Walter Brownlow, ed. "Alabama in the 1830's; As Recorded By British Travelers." *Birmingham Southern College Bulletin* 31 (December 1938).

Prunty, Merle, Jr. "Recent Quantitative Changes in the Cotton Regions of the Southeastern States." *Economic Geography* 27 (July 1951): 189–208.

———. "The Renaissance of the Southern Plantation." *Geographical Review* 45 (October 1955): 459–91.

Rea, Robert R. "Planters and Plantations in British West Florida." *Alabama Review* 29 (July 1976): 220–35.

Rifkin, Jeremy. *Beyond Beef: The Rise and Fall of the Cattle Culture.* New York: Penguin, 1992.

Riley, Benjamin Franklin. *History of Conecuh County, Alabama.* 1881. Reprint. Blue Hill, Maine: Weekly Packet, 1964.

Rogers, William Warren. *The One-Gallused Rebellion: Agrarianism in Alabama, 1865–1896.* Baton Rouge: Louisiana State University Press, 1970.

Rogers, William Warren, Robert David Ward, Leah Rawls Atkins, and Wayne Flynt. *Alabama: The History of a Deep South State.* Tuscaloosa: University of Alabama Press, 1994.

Saloutos, Theodore. *The American Farmer and the New Deal.* Ames: Iowa State University Press, 1982.

———. *Farmer Movements in the South, 1865–1933.* Berkeley: University of California Press, 1960.

Schafer, Elizabeth Diane. "Reveille for Professionalism: Alabama Veterinary Medical Association, 1907–1952." M.A. Thesis, Auburn University, 1988.

Schwabe, Calvin W. "Man of the New South—Charles Allen Cary: Auburn's Rural Health Pioneer." Unpublished paper presented at the centennial celebration of Cary's arrival at Auburn, November 1992. Copies in possession of the author.

Scott, Roy V. *The Reluctant Farmer: The Rise of Agricultural Extension to 1914.* Urbana: University of Illinois Press, 1970.

Sellers, James Benson. *Slavery in Alabama.* University, Alabama: University of Alabama Press, 1950.

Simms, Jack. "Charles Allen Cary, the Citizen." Unpublished paper presented at the centennial celebration of Cary's arrival at Auburn, November 1992. Copies in possession of the author.

Sisk, Glenn N. "Alabama's Black Belt: A Social History, 1875–1917." Ph.D. Diss., Duke University, Durham, North Carolina, 1951.

Stepp, Wendell H., and Pamela Ann Stepp. *Dothan: A Pictorial History.* Norfolk, Virginia: The Donning Co., 1984.

Stickney, Hazel. "The Conversion from Cotton to Cattle Economy in the Alabama Black Belt, 1930–1960." Ph.D. Diss., Clark University, Worcester, Massachusetts, 1961.

Taylor, Paul Wayne. "Mobile: 1818–1859 as Her Newspapers Pictured Her." M.A. Thesis, University of Alabama, 1951.

Tindall, George B. *The Emergence of the New South, 1913–1945.* Baton Rouge: Louisiana State University Press, 1967.

Tower, J. Allen. "Alabama's Shifting Cotton Belt." *Alabama Review* 1 (January 1948): 27–38.

Watkins, C. Robert. "General Wilson's Raid through Alabama and Georgia in 1865." M.A. Thesis, Auburn University, 1959.

Watson, Fred S. *Coffee Grounds: A History of Coffee County, Alabama, 1841–1970.* Anniston, Alabama: Higginbotham, 1970.

Wiebe, Robert H. *The Search for Order, 1877–1920.* The Making of America Series. New York: Hill and Wang, 1967.

INDEX

Abbeville, Ala., 36
Aberdeen-Angus cattle. *See* Angus cattle
Adams, James L., 170
Ad valorem taxes, 135
Agricultural Adjustment Administration, 92, 96, 98, 99, 105, 110, 116, 139, 146
Agricultural and Mechanical College of Alabama. *See* Auburn University
Agricultural Committee of the Alabama Bankers' Association, 97
Agricultural Credit Bank, 97
Agricultural progressivism, 43–44, 61–62, 63, 64, 74, 77, 92, 134
Agricultural Wheel, 51
Alabama Agricultural Experiment Station, 56, 57–58, 64, 65, 70, 100, 103, 108, 127, 128, 138, 147; beef cattle experiments, 131, 147, 148; bulletins of, 58, 110; crop experiments, 58, 88–89; pasture and forage experiments, 128, 147; system of substations established, 90
Alabama Agricultural Statistics Service, 118
Alabama Angus Breeders' Association, 69, 73, 79, 106
Alabama Beef Cattle Show and Sale, 119
Alabama Beef Council, 136
Alabama Blue Book and Social Register, 68
Alabama Brahman Association, 149
Alabama Cattleman, 136
Alabama Cattlemen's Association, 109, 120–22, 127, 128, 132–37, 139, 140, 141, 147, 150–51, 158, 162; beef promotion by, 136–37, 150, 163; organizational structure of, 135–36; political influence of, 133–34, 135, 148, 150
Alabama Chamber of Commerce, 116, 119, 121–22, 132, 133, 136
Alabama Charolais Association, 149
Alabama Constitution (1901), 53
Alabama Cow Belles Association, 136

Alabama Department of Agriculture and Industries, 71
Alabama Farm Bureau Cotton Association, 68
Alabama Flour Mills, 146
Alabama Hereford Breeders Association, 69, 73–74, 79, 106
Alabama Jersey Cattle Club, 83
Alabama legislature, Acts of, 37, 40, 50, 54, 56, 61, 90, 106, 112, 126, 127, 132, 133–34, 135, 136
Alabama Lime Belt Feeder Calf Association, 93, 94, 95
Alabama Live Stock Association, 60, 67, 80
Alabama Livestock Growers Association, 106, 120
Alabama National Bank, 68
Alabama Polled Hereford Association, 149
Alabama Polytechnic Institute. *See* Auburn University
Alabama Power Company, 67
Alabama River, 3, 5, 12, 21, 140
Alabama State Agricultural Center, 135
Alabama State Agricultural Society, 24, 56, 57
Alabama State Board of Agriculture, 106
Alabama State Council of Defense, 55, 78
Alabama State Fair, 72, 89. *See also* Birmingham State Fair; Livestock shows and fairs
Alabama State Live Stock Sanitary Board, 60, 61
Alabama State Milk Control Board, 122
Alabama Supreme Court, 53, 54, 55
Alabamas, 4
Albertville, Ala., 155
Alderney cattle. *See* Jersey cattle
Alexander, Whit, 52
Alexander City, Ala., 83
Alfalfa, 65, 159
Allen, L. J., 53
Allis-Chalmers Roto-Baler, 161
Allison, E. F., 58

· 207 ·

INDEX

American Cattlemen's Association, 136
American Cotton Planter and Soil of the South, 24, 49
American Dairy Association of Alabama, 125
American International Charolais Association, 150
American Review of Reviews, 85
American Short-Horn Herd Book, 73
Andalusia, Ala., 65, 81, 125
Angel, James, 48
Anglo-Americans: herders, 15; settling Alabama, 5, 7, 9, 10, 16
Angus cattle, 58, 67, 68, 69, 71, 72, 73, 78, 79, 93, 109, 116, 119, 130, 135, 138, 140, 147, 148, 183 (n. 34), 186 (n. 74)
Anniston, Ala., 101
Appalachian region, 151; cattle auctions in, 132, 155; Civil War in, 35–36; conflicts over closing of range in, 52–53; cooperative extension efforts in, 130; cotton raising in, 46, 49; dairy farming in, 123, 125, 152, 153, 154; depletion of cattle in, 41; drovers in, 46; increase of cattle numbers in, 119, 152, 153–57, 162; lack of commercial cattle industry in, 110, 111, 116, 126; open-range herding in, 17, 20, 25, 28–29, 31, 46; settlement of, 16; small and part-time farmers in, 154–55, 157, 158, 164; tick eradication in, 61–63
Armistead, G. G., 22
Armour and Company, 117, 146
Armstrong, John, 140, 149
Army of Mississippi, 36
Arrington, Robert Goldthwaite, 67, 72, 96, 170
Ashville, Ala., 155
Athens, Ala., 20, 34, 109
Athey, T. Whit, Jr., 89, 98, 104, 132, 170
Athey, T. Whit, Sr., 89–90
Atlanta, Georgia, 23, 38, 80, 94, 131
Atlantic Coast Line Railroad, 94
Auburn, Ala., 57, 58
Auburn University, 44, 55, 58, 59, 64, 68, 79, 81, 83, 85, 93, 100, 104, 108, 121, 127, 131, 132, 136, 140, 146, 147, 148, 149, 150, 163, 164, 184–85 (n. 54); Animal Husbandry Department, 185 (n. 54); College of Veterinary Medicine, 59
Auctions, 92, 106–7, 127, 131, 132, 134, 155. *See also* Marketing cattle

AU Lean, 163
Autauga County, 23, 66, 107, 126, 139, 140
Ayershire cattle, 24

Bahia grass, 130
Bailey, R. Y., 101, 102–3
Bailey, W. J., 154
Baker, K. G., 90–91, 103–4, 120, 130
Baldwin, William O., 23
Baldwin County, 12, 26, 45, 46, 48, 61, 95, 111, 118, 148
Bang's disease. *See* Brucellosis
Bankhead-Jones Act, 104
Banks, A. D., 36, 38
Barber, George H., 124
Barber, George W., 124
Barber, Warren, 124
Barber Dairies, 124, 125
Barbour County, 21, 99, 118
Barrow, John, 18
Bartram, William, 8–9
Bayer, S. D., 95
Beane, J. C., 83
Beard, Richard, Sr., 170
Beef, 14, 20, 38, 46, 48, 145; among Native Americans, 2; barreled, 12; Confederate ration of, 181 (n. 72); consumption of by antebellum southerners, 27, 28; in Confederate commissary, 36, 41; inspection of, 60
Beef cattle course, 185 (n. 54)
Beef Cattle Field Day and Bull Sale, 147, 148
Beef Cattle Improvement Association, 147, 150
Beef Cattle Performance Testing Program, 149
Beef checkoff, 136–37, 150, 163
Beef Herd Improvement Association, 130
Beef import quotas, 150
Beefmaster cattle, 149
Beef Promotion Bill, 136
Bell, N. J., 74, 79, 80, 97
Belle Mina, Ala., 122
Bermuda grass, 71, 130
Beyond Beef: The Rise and Fall of the Cattle Culture, 163
Bibb County, 181 (n. 74)
Biloxi Bay, 3
Birmingham, 80, 81, 82, 83, 84, 94, 119, 122, 123, 124, 125, 133, 140, 144, 194 (n. 54)
Birmingham News, 103, 109
Birmingham State Fair, 106, 108. *See also*

INDEX

Alabama State Fair; Livestock shows and fairs
Black Belt, 7, 18, 29, 30, 31, 41, 43, 45, 88, 90, 97, 107, 108, 116, 121; agricultural research in, 56, 58, 65, 89; antebellum breed improvement in, 21, 22, 23–25; boll weevil in, 70–71, 77, 141, 168; cattle auctions in, 106, 131; closing of the open range in, 50; conversion from cotton to cattle in, 66–67, 70–71, 86, 87, 89, 103, 108, 110, 138–40, 141; cooperative extension efforts in, 93–94, 104; cotton planting in, 9, 47, 65, 70–71, 76, 111, 139, 140, 157; dairy farming in, 65, 107, 123, 124; decline of cattle numbers in, 152, 156, 157, 162; development of midwestern-style cattle industry in, 65–67, 70–71, 72, 73, 77, 87, 92, 93, 95, 112, 138, 140, 164; drovers in, 47; feeder cattle in, 146; herders in, 15; impressment in, 38; increase of cattle numbers in, 26, 28, 48, 87, 119, 126, 137, 141; influence in Alabama Cattlemen's Association, 120, 132, 133, 137, 151, 162, 194 (n. 54); large cattle farms in, 158; midwestern immigration into, 69–70, 140; opposition to beef check-off in, 136; planter-livestock raisers in, 22, 24, 47, 48, 66–68, 70–71, 76, 88, 107, 108, 120, 138, 139; production credit system in, 97; settlement of, 16; tick eradication in, 59, 60, 65; twentieth-century breed improvement in, 72, 73, 78, 79–80, 93, 140, 148
Black Belt Creamery, 123, 125
Black Belt Feeder Calf Sales, 116–17
Black Belt Feeder Cattle Producers Association, 116, 117
Blackbelt Livestock Association, 60
Black Belt Stock Yard Corporation, 140
Black Belt Substation, 90–91, 92, 103–4, 120, 140
Black farmers and laborers, 77, 85, 113, 138, 139–40, 156
Black medic, 91
Black outmigration, 77, 84–86, 138
Blackwood, Houston, 118
Blevins, Pauline, 155
Blevins, Noel, 155
Blount County, 118, 119, 153, 155
Blue Bonnett, 123
Bluegrass, 109
Bluestem, 130
Blythe, Greg, 171
Boaz, Ala., 123, 154
Boll weevil, 58, 60, 64–65, 70–71, 72, 78, 84, 86, 186 (n. 68)
Bonner, James C., 22
Booth, Joseph, 27
Bourbon Stockyards, 80, 81
Bovine tuberculosis, 60
Braford cattle, 149
Bragg, Braxton, 32
Brahman cattle, 131, 148, 149, 195 (n. 8)
Branding, 9, 19, 106, 134
Brangus cattle, 149
Breeder's Gazette, 87
Breen, W. P., 133, 170
Brice, S. A., 186 (n. 73)
British: cattle breeds, 4, 22, 23, 44, 79, 112, 113, 131, 148, 149, 150; in the Carolinas, 3, 4; in West Florida, 5
Brown, Thomas B., 23
Brown, W. M., 170
Brown Swiss cattle, 123
Browne, Montfort, 4
Brucellosis, 127, 147
Brumfield, Joseph, 46
Buck and Sandy Creeks Conservation Project, 101
Buell, Don Carlos, 33
Bullock County, 71, 93, 107, 117, 157, 168, 185 (n. 57)
Bully (Creek herder), 9
Bureau of Animal Industry, 59, 63, 89
Bureau of Entomology, 100
Burgreen, Carl Edward, 109
Burnett, Sam, 154
Burns, Lucien Powell, 68
Burton, Glenn, 130
Burton, W. J., 83
Butler County, 184 (n. 41)
Bynum, William, 20
Byrd, David, 17

Cahaba River, 37
Caley family, 69, 140
Caley peas, 91
Camden, Ala., 116
Campbell, Christiana McFayden, 104
Campbell, Thomas, 85
Camp Hill, Ala., 123

INDEX

Canebrake Agricultural District, 50
Canebrake Agricultural Experiment Station, 56, 57, 65
Canebrake Cow Testing Association, 83
Canebrakes, 6, 8, 10, 15, 20, 111
Carney, James, 46
Carolinas, 17. *See also* North Carolina; South Carolina
Carrithers, W. D., 80
Cary, Charles Allen, 59–60, 61, 63, 79, 91
Cattle. *See* individual breeds and types
Cattle prices, 39, 77, 86, 87–88, 89, 99, 111, 118, 120, 127, 128, 153, 156, 158–59, 163. *See also* Marketing cattle
Cattle raising. *See* Appalachian region; Black Belt; Piedmont; Piney woods region; Tennessee Valley; Wiregrass
Cattle rustling, 36–37, 106, 134, 135
Cattle tick, 74; eradication of, 58–63, 65. *See also* Dipping vats
Cawthon, William, 17
Cedartown, Georgia, 123
Central Alabama Soil Conservation District, 103
Chambers County, 83, 119, 194 (n. 53)
Charolais cattle, 149–50
Chattahoochee River, 6
Chattanooga, Tennessee, 30, 35, 155
Cherokee County, 36, 109, 121, 154
Cherokees, 4, 16
Cherry, F. L., 17
Chianina cattle, 149
Chickasaw Land Company, 53
Choctaw County, 103, 123, 134, 181 (n. 74)
Choctaws, 3, 16
Cincinnati Abatoir Company, 81
Civil War: in Alabama, 29–42; impact on cattle industry, 30–32, 41–42, 45
Civil Works Administration, 101
Civilian Conservation Corps, 101, 138
Claiborne, J. F. H., 18–19
Clanton, James H., 35
Clarke County, 17, 63, 95, 118, 123, 134, 181 (n. 74)
Clay, C. C., 78, 81
Clay, Governor C. C., 34
Clay County, 49, 100, 110, 156, 183 (n. 26)
Clayton, Preston, 170
Clean Water Act, 163
Clear Creek Lumber Company, 53
Cleburne County, 110, 194 (n. 53)

Clemen, Rudolf Alexander, 65
Cloud, Noah, 23, 24, 49, 77
Coastal Plains, 131
Cobb, Oscar, 72, 74, 80, 186 (n. 74)
Coe, Tim, 171
Coffee County, 17, 49, 111, 118
Coker, Wilson, 148
Colbert County, 111, 118, 127, 181 (n. 74)
Columbus, Georgia, 9, 17
Commission buyers, 80, 108, 131. *See also* Marketing cattle
Commissioner of Agriculture and Industries, 134
Compton, A. W., 171
Conecuh County, 17, 26
Confederacy: beef ration in, 181 (n. 72); beef in commissary, 36, 41
Confederate Congress, 40
Conveyor (hay elevator), 161
Conway, Charles, 12
Cooperative Extension Service, 58, 91, 93, 121, 147, 150; black agents, 78–79; cooperation with Alabama Cattlemen's Association, 128, 132; cooperation with Farm Bureau, 105; Department of Animal Husbandry, 83; efforts to expand beef cattle raising, 78–79, 81, 89, 94, 95, 104, 106, 109, 110, 117, 118, 119, 127, 130, 138, 152, 153, 154; efforts to expand dairy farming, 83; involvement in New Deal programs, 92, 96, 100, 103, 104, 116
Coosa County, 30, 62, 110, 154
Coosa River, 5, 6, 9
Cort, Julio, Jr., 170
Cotton Belt Railroad, 91
Cotton planting and planters, 12, 13, 21, 51, 56, 72, 76, 86, 88, 138, 139, 154; boll weevil's effect on, 64–65, 70–71; cattle raising among, 15, 21–22, 23–24, 47, 77, 107, 109, 113; consumption of beef by, 28; decline of, 110, 114, 118, 125, 152; expansion of, 14, 16, 18, 25, 26, 27, 29, 31, 42, 44, 45, 48–49; mechanization of harvest, 153, 156; New Deal's effects on, 96–97; settlement by, 9, 15
Country buyers, 80, 131, 132, 193 (n. 41). *See also* Marketing cattle
Courtland, Ala., 52
Covington County, 18, 26, 27, 36, 37, 46, 49, 55, 62, 111, 112, 118
Cow-calf farming, 144, 147

INDEX

Cowhides, 2, 4, 12, 28, 128; among Native Americans, 2; whips made of, 9
Cow peas, 108
Cowpens: plantations, 4, 7, 10–11; stockades, 9, 17, 22, 25
Crabb, Morton C., 69
Cracker cattle. *See* Native cattle
Cravy, James, 111–12
Cravy, N. B., 55
Crawford, Joe R., 170
Crawford, M. L., 89, 90
Crawford, Roy J., 90, 96
Credit. *See* Production credit system
Creeks, 2, 4, 9, 16, 17; cattle raising among, 6; conflicts with white settlers, 5, 7
Crenshaw County, 89, 104
Crimson clover, 91, 101, 103, 108, 109, 130, 159
Croom, Isaac, 23, 24, 49, 77
Crossbreeding, 109, 131, 147, 148, 149, 150
Croxton, John, 37
Cullman, Ala., 155
Cullman County, 49, 100, 154, 155, 156, 157
Cullman Democrat, 63
Cullman Stockyards, 155
Curtis, Frank, 78
Cusseta, Treaty of, 16

Dadeville, Ala., 101
Dadeville Soil Conservation Demonstration Project, 102
Dairy, 9, 23; conversion from, to beef, 107; creameries, 83, 122, 123, 125; experiment station research in, 56, 122; farming as integral to subsistence, 45, 47–48; farming in Appalachia, 46, 83, 152, 153, 154; farming in Black Belt, 23, 65, 72, 81, 82, 103, 122, 139; farming in Tennessee Valley, 83, 118; growth of during World War I, 81–83; growth of during World War II, 122–23; industry, 123–25; milk inspection laws, 60, 122, 124; neglect of in antebellum South, 8, 17; urban demand for products, 44, 81; whole milk market, 122, 123, 125
Dairy Fresh Company, 125
Dale County, 49, 111
Dallas County, 18, 28, 50, 65, 67, 68, 69, 70, 71, 137, 138–39, 140, 156, 157, 158, 168, 184 (n. 41), 185 (n. 57)
Dallis grass, 91, 103, 138
Dancing Rabbit Creek, Treaty of, 16
Daniel, Pete, 64, 79

Darr, Francis, 33
Dauphin Island, 1, 3, 4, 8
Davis, A. C., 81
Davis, Hugh, 21
Davis, Jefferson, 38, 39
Davis, P. O., 85–86, 100, 104, 121, 128, 132
D'Bienville, Jean-Baptiste Le Moyne, 3
De Graveline, Jean-Baptiste Boudreau, 3
D'Iberville, Pierre Le Moyne, 3
De León, Ponce, 1
De Luna, Don Tristan, 2
De Soto, Hernando, 1
De Velasco, Luis, 2
Dearmon, John, 18
Debter, Glynn, 171
Decatur, Ala., 37, 146
Deerhead Cove, Ala., 155
DeKalb County, 20, 21, 29, 36, 45, 46, 49, 54, 62, 155, 156, 157
Demopolis, Ala., 24, 38, 49, 69, 78, 79, 80, 81, 106, 116, 117, 119, 120, 121, 131, 140, 149
Demopolis Production Credit Association, 117
Demopolis Stockyards, 117
Dent, John Horry, 21, 22
Derby, F. I., 58, 72, 74, 186 (n. 74)
Devon cattle, 23, 24, 25, 62
Dipping vats, 61, 62; dynamiting of, 63. *See also* Cattle tick: eradication of
Dixie Dairies, 125
Dodge, Grenville M., 34
Donaldson, Ronny, 171
Doran, Michael F., 5
Dothan, Ala., 17, 83, 106, 125
Drake, K. Stanley, 170
Drought-cattle relief program, 92, 96, 98–99
Drovers, 14, 20, 45, 46–47
Duggar, J. F., 78
Dunaway, John Eugene, 67, 69, 74, 79
Duncan, L. N., 105
Durham cattle. *See* Shorthorn cattle
Durnford, Elias, 4

Earle (doctor), 17
Earth Day, 163
Easley, Colonel, 7
East Alabama Cattlemen's Association, 117
Eastern Charolais and Charbray Association, 149
Easy Way Baler Lodr, 161

INDEX

Eaton, William Hunt, 83
Edwards, G. M., 117, 148
Eicher and Meninger (packers), 117
Elk River, 33, 109
Ellis, Ned, 171
Ellis farm, 139
Elmore County, 53, 119, 184 (n. 41)
Elsberry, William Edward, Jr., 68
Elyton, Ala., 37
Embry, F. H., 80
Embry, G. W., 80
Emergency Conservation Works, 101
Emistisiugo (Creek chief), 5, 7
Engerman, Stanley L., 22
Enterprise, Ala., 125
Environmentalism, 162, 163
Environmental Protection Agency, 163
Epes, Ala., 106, 116, 117, 131
Escambia County, 49, 61, 134
Eslava, Don Miguel, 12
Etowah County, 62, 181 (n. 74)
Eufaula, Ala., 128
Eutaw, Ala., 131
Ewell, Benjamin S., 35
Exotic cattle breeds, 143, 147, 148, 149

Farm Animal Reform Movement, 162
Farm Bureau, 92, 93, 95, 104–5, 109, 121, 136, 141
Farm Bureau Commission Firm, 93
Farm Credit Administration, 97
Farmers' Alliance, 51
Farmers' Institute, 64, 184 (n. 54)
Farmhand Bale Accumulator, 161
Federal Reserve Bank of Birmingham, 67, 140
Federal Surplus Relief Corporation, 98
Feeder cattle, 138, 140, 144, 145–46
Feedlots, 146–47, 163
Fertilizer, 119; for cotton, 48–49, 56; for pasture and forage crops, 91, 100, 101, 102, 108, 109, 130
Fescue, 130, 154, 159
Fields farm, 138
Finishing cattle. *See* Feeder cattle
Fite, Gilbert C., 105, 115, 130
Fitzpatrick, L. D., 171
Fleming, Walter L., 33
Flint River, 5, 6
Florence, Ala., 73, 121, 124

Florida, 3, 8, 17, 18, 32, 36, 37, 42, 94, 98, 107, 110; Spanish in, 1–2
Fogel, Robert W., 22
Foremost Dairies, 124
Forrester, W. J., 83
Fort Jackson, Treaty of, 16
Fort Payne, Ala., 73, 155
Fort Payne Stockyard, 155
Fort Deposit, Ala., 116
4-H, 83, 128, 130
4-H beef calf clubs, 93, 104, 106, 110, 128
4-H beef calf show, 106
4-H Jersey cattle clubs, 83
Franklin County, 33
French: settlement, 3; cattle breeds, 149, 150
Fuller, Luther, 119–20, 121, 135

Gadsden, Ala., 35, 63, 128
Gallion, Ala., 69
Garner, James, 163
Garrett Coliseum, 135
Gastonburg, Ala., 67
Gates, Paul, 32
Gelbvieh cattle, 149
General Education Board, 123
General Orders No. 81, 34
Geneva County, 18, 36, 49, 62, 118, 152
Genovese, Eugene D., 14
Georgia, 2, 7, 9, 15, 16, 17, 21, 28, 35, 37, 38, 50, 99
Georgia Agricultural Experiment Station, 130
Gilmer, Francis Merriweather, 22
Goode, Robert James, Jr., 67, 79, 95
Gosse, Philip Henry, 18
Gossett, J. W., 150
Grady, Ala., 89
Grand Bay, Ala., 83
Grange, 51, 56, 188 (n. 18)
Gray, Henry B., III, 170
Gray, Lewis C., 16, 23
Green Bay, Ala., 55
Greene County, 50, 95, 126, 133, 157, 168, 185 (n. 57)
Greene County, Mississippi, 19
Greensboro, Ala., 23, 95, 124, 125
Greenville, Ala., 101, 136
Gregory, William, 6
Gregory, William H. "Mutt," 106, 121, 127, 132, 135
Greil, Gaston J., 68

INDEX

Grierson, Robert, 6
Griffith, E. P., 117
Grimes, J. C., 89
Grossman, James R., 85, 86
Guernsey cattle, 123, 153
Guice, John D. W., 12, 13, 17, 25
Gulf Coast Substation, 90
Guntersville, Ala., 155
Guthrie, Mayo, 154
Guy, W. W., 35–36

Hackney, Sheldon, 53, 182 (n. 21)
Hahn, Steven, 50–51
Haigler Brothers, 95
Hale, E. E., 95
Hale County, 69, 70, 71, 87, 124, 140, 156, 157, 168, 185 (n. 57)
Hall, Bolling, 23
Hall, M. W., 83
Hall, M. W., Jr., 107
Hamburg, Ala., 90
Ham Wilson Livestock Arena, 148
Hansen Pure Milk Company, 124
Hardee, Phil, 171
Hardie, William, 83
Hart, Allen, 18
Hart, Dennis, 46
Hart, James E., 170
Hartley, A. C., 107
Hartselle, Ala., 190 (n. 59)
Hatch Act, 56, 57
Haujo, Chief Toolkaubatche, 6
Havana, Cuba, 3
Hawkins, Benjamin, 5–7
Hawkins' Bill, 56
Hay, 70, 71, 101, 102, 119, 139, 152, 159; balers, 139, 160–61, 162; harvest as social event, 159; round balers, 162
Hay elevator, 161
Hay Growers Association, 70
Hayneville, Ala., 95
Health consciousness, 162–63
Hearin, Jesse, 97
Heck, James, 46
Helms, Bob, 171
Henderson, Charles, 60
Henderson, J. Bruce, 120, 136–37, 148, 170
Henderson, W. G., 79
Hengst, F. D., 73
Henley, O. J., 133, 170, 194 (n. 54)

Henry County, 18, 111, 118, 181 (n. 74)
Herders, 14, 15, 20, 29, 35, 46, 75, 135
Herding. *See* Open-range herding
Hereford cattle, 23, 67, 68, 71, 72, 73, 78, 79, 89, 93, 94, 106, 108, 109, 116, 119, 130, 135, 139, 140, 147, 148, 149, 185 (n. 62). *See also* Polled Hereford cattle
Hereford Journal, 73
Herrin, Stephen W., 21, 26
Herzfield, Harry, 83
Heston balers, 162
Hickey, Daniel, 4
Hickson, James, 46
Hill country. *See* Appalachian region
Hilliard, Sam Bowers, 21, 23, 27, 28, 41
Hinds, Warren E., 64
Hobby farming, 68, 72
Hodgson, Adam, 17, 20
Hog raising, 1, 6, 10, 18, 32, 89, 90, 109, 125; among Native Americans, 16; cooperative extension efforts with, 104; in Appalachia, 10, 15, 20, 21, 26, 31, 45, 158; in piney woods, 18, 27, 45; in Tennessee Valley, 20
Holladay, Fred, 103
Holladay, Ronnie, 171
Holley, Alfred, 37
Hollis, Dan W., 121
Holstein cattle, 72, 83, 107, 123, 154
Homewood, Ala., 124
Hood Farm, 82
Hooper Stock Yards, 149
Hop clover, 109
Horseshoe Bend, Battle of, 16
Horton, J. Ed, Jr., 170
Houghton, H. S., 68, 72
Houston County, 18, 119, 152
Howard, Bo, 141, 170
Huddleston, Nolan, 105
Huddleston, W. E., 107
Huffman, Dale, 163
Huntsville, Ala., 32, 34, 73, 80, 83, 109

Impressment, 32, 33, 34, 35, 36, 38–40
Indian Removal Act of 1830, 16
Indian Service, 98
Inner frontiers, 14, 15, 23, 25, 31, 45, 48
International Harvester Model 50-T baler, 160
Iowa, 66, 89, 149
Iowa State College, 59, 66
Irving, T. B., 69

INDEX

Jackson, Ala., 95
Jackson County, 33, 34, 157, 181 (n. 74), 184 (n. 41)
Jackson Land and Lumber Company, 112
Jasper, Ala., 52, 53, 155
Jefferson, Thomas, 7
Jefferson County, 81, 124
Jemison, Robert, 82, 94
Jenkins, U. C., 89, 95, 96, 120
Jersey cattle, 24, 56, 72, 79, 82, 83, 94, 123, 124, 153, 154
John Deere balers, 161, 162
Johnson, Bill, 171
Johnson, Harold, 170
Johnson, William H., Jr., 104, 134
Johnson grass, 70, 71, 91, 107, 108
Johnston (Washington County herder-planter), 18
Johnston, Joseph E., 35, 38
Joiner, Orrin, 53
Jones, Carol, 139–40
Jones, Edwin, 130
Jones, Madison, 95
Jones, Raymond B., 170
Jones, Walter B., 136
Jordan, Mortimer H., IV, 133, 135, 170, 194 (n. 54)
Jordan, Terry G., 2, 3, 13, 25
Judson College, 108
Judy, M. A., 79

Kennett, William L., 80
Kennett-Murray, 80–81
Kentucky, 68, 69, 80, 94, 117, 186 (n. 66)
Kernachan, J. S., 58, 73
Kilby, Thomas E., 63
Killian, Dr. George S., 170
Killough, Forrest, 170
Kilpatrick, Ala., 155
Kimbrough, W. B., 47
King Ranch (Texas), 148
Kinzer, R. J., 73
Kirkwood Plantation, 88–89
Knapp, Seaman A., 64
Kraft Cheese, 123
Kudzu, 102–3, 108, 118

Laker, Peter, 18
Lambert, C. Roger, 98
Lambert, J. E., 149, 170
Lambert, Joe, 67

Lambert, R., 95
Lambert Meats Laboratory, 147
Lane, Cecil, 170
Lane, E. P., 146
Lang, k. d., 162
Lanier, W. R., 170
Larkin farm, 138
Lauderdale County, 22, 29, 33, 47, 48, 58, 73, 94, 104, 119, 184 (n. 41)
Lawrence County, 33, 52, 119
Leaksville, Mississippi, 19
Leather, 12, 20
Lee County, 51, 83, 86
Lespedeza, 103, 109, 118
Lewis, V. W., 94
Libby, McNeil and Libby, 124
Limestone County, 33, 59, 94, 95, 109, 111, 126, 184 (n. 41), 186 (n. 74)
Limousin cattle, 149
Linden, Ala., 106, 131
Link, William A., 62
Linton, Daniel, Jr., 146
Little River, 6
Livestock Conservation Campaign, 55
Livestock Producers Association, 95, 96
Livestock shows and fairs, 24, 67, 72, 106, 108, 119, 133, 186 (n. 73)
Livingston, Ala., 71, 131
Loachapoka, Ala., 37
Lockhart, Ala., 73
Longhorn cattle, 135, 163
Louisiana, 28, 150
Louisville, Kentucky, 33, 80, 81, 95, 108
Louisville and Nashville Railroad, 35, 55
Lowe, Robert J., 120–21, 170
Lowndes County, 22, 28, 68, 80, 94, 103, 126, 137, 148, 156, 157, 158, 168, 181 (n. 74), 184 (n. 41), 185 (n. 57), 188 (n. 32)

McBeath, O. G., 146
McConnico, Stonewall, 67, 94, 95
McConnico, William Washington, 67
McDonald, Forrest, 14, 28, 31, 42, 45
McEwen, T. A., 154
McGillivray, Alexander, 9
McGillivray, Daniel, 6
McKenzie, Robert H., 37
McLean burger, 163
McMillan, Elizabeth, 46
Macnack, Sam, 6
McNeill, Frank, 47

INDEX

Macon County, 83, 107
Macon-Lee Jersey Bull Association, 83
McQueen, Peter, 6
McWhiney, Grady, 14, 28, 31, 42, 45
McWilliams, A. L., 118
Madison County, 33, 47, 59, 109, 121, 126, 133, 150, 184 (n. 41)
Maldonado, Don Diego, 1
Maples, Billy, 171
Maples, Joe, 109
Maples, Malcolm "Mack," 109, 136, 170
Marbilou Ranch, 140
Marengo County, 22, 24, 28, 47, 48, 49, 50, 53, 78, 83, 88, 133, 141, 157, 158, 168, 181 (n. 74), 184 (n. 41), 185 (n. 57)
Marengo Farms, 80
Marion, Ala., 101, 131
Marion County, 181 (n. 74)
Marion Institute, 67
Marion Junction, Ala., 70, 72, 90, 186 (n. 73)
Marketing cattle, 46, 80, 106, 131–32, 134; auctions, 92, 106–7, 127, 131, 132, 134, 155; commission buyers, 80, 108, 131; country buyers, 80, 131, 132, 193 (n. 41); terminal markets, 80, 131
Marshall County, 61, 155
Martineau, Harriet, 9
Matthews, Ala., 107
Mealing, Ed, 103
Mease, Edward, 8
Meat packing. *See* Packing industry
"Meat Stinks," 162
Medlock, Olin, 101
Melish, John, 9
Midway, Ala., 83
Midwestern-style cattle industry, 65–66, 69–70, 71, 72–74, 76, 95, 112, 164
Miller's Ferry, Ala., 120
Milner, John T., 30
Mississippi, 18, 28, 35, 38, 110
Missouri, 66, 93, 94
Mitchel, Ormsby, 32, 33, 34
Mobile, 4, 24, 32, 36, 38, 55, 72, 72, 81, 122, 125, 133, 141; cattle market in, 8, 11, 17, 19–20, 28, 46; French in, 3
Mobile Bay, 5
Mobile & Ohio Railroad, 55
Mobile County, 12, 53, 63, 64, 94, 111, 118, 133, 134, 141
Monroe County, 109, 111, 181 (n. 74), 184 (n. 41)

Monroeville, Ala., 106
Montevallo, Ala., 37, 154
Montgomery, Ala., 9, 23, 24, 38, 67, 68, 72, 79, 80, 81, 83, 84, 89, 92, 93, 94, 97, 99, 100, 106, 108, 116, 117, 121, 122, 125, 131, 133, 134, 135, 136, 140, 149, 150, 164
Montgomery Abatoir, 81
Montgomery Advertiser, 34, 60, 69, 71, 73
Montgomery and Pensacola Railroad, 36
Montgomery and West Point Railroad, 37
Montgomery Cattlemen's Club, 60
Montgomery County, 22, 68, 69, 71, 89, 107, 110, 119, 137, 140, 141, 146, 148, 156, 157, 168, 184 (n. 41), 185 (n. 57)
Montgomery Livestock Association, 68, 72
Montgomery Live Stock Show, 72
Montgomery Live Stock Show Association, 72
Montgomery Production Credit Association, 68, 72
Moore, A. B., 186 (n. 73)
Moore, J. D., 92–95, 96, 99, 110, 188 (n. 32)
Moore, Richard, 46, 48
Morgan County, 20, 33, 103, 121, 126, 155, 190 (n. 59)
Morrill Land-Grant Act, 55
Morris, Ala., 155
Morrison, J. Lem, 124–25
Mountain Eagle (Jasper), 52
Murphy, W. W., 71

Nashville, Tennessee, 33, 95
National Price Administration, 158
Native cattle, 10, 23, 28, 79, 135
Native Americans, 2, 3, 8. *See also* Alabamas; Cherokees; Choctaws; Creeks
Neoplantation, 139
Nestle Company, 125
New Deal, 92, 96, 97, 100, 103, 104, 116
New Echota, Treaty of, 16
New Holland: Bale Thrower, 161; Model 66 Bailer, 161; Stackliner, 162
Newman, Dr. A. C., Jr., 170
Newman, James, 56
New South creed, 55–56, 61
New Orleans, Louisiana, 80
New Orleans Agricultural Credit Bank, 107
1901 Alabama Constitution, 53
Niven, L. A., 87
North Carolina, 59, 94, 117
Northrop, Lucius, 38–39

Oakman, Ala., 52
Oats, 91, 108
Ocfuskee, 6
Odom, Bessie, 154
Oklahoma, 106, 145
Opelika, Ala., 37, 38, 51, 56
Opelika Times, 51, 52
Open-range herding, 14, 21, 24, 29, 42, 45, 61, 63, 77, 111–12; demise of, 24, 48, 49–55, 112, 126, 134. *See also* Drovers; Herders; Piney Woods region: open-range herding in
Orchard grass, 109
Orrville, Ala., 67
Otto, John Solomon, 13, 26, 32, 41
Overton, Hugh, 118
Owsley, Frank L., 13, 15, 16–17, 25, 26, 28
Oxford, Ala., 35, 83
O'Neal, Edward A., 104–5

Packing industry, 65, 80, 99, 108, 117, 127, 131, 144, 145
Paine, J. J., 83
Park, John, 21
Parrish, Keith, 155
Part-time farming, 143, 147, 154–55, 157, 158
Pate, Harold E., 170
Peanut farming, 90, 110, 118, 126
Pembroke, Kentucky, 130
Pensacola, Florida, 2, 4, 6, 9, 11, 17
People for the Ethical Treatment of Animals, 162
Perdido River, 8
Perry County, 50, 56, 71, 82, 90, 157, 168, 184 (n. 41), 185 (n. 57)
Peters, John, 22
Peters, Richard, 23
Phillips, G. B., 94
Pickens County, 184 (n. 41)
Piedmont, 7, 151; closing of the open range in, 51; cotton planting in, 9, 11; dairy farming in, 123; herders in, 15; increase of cattle numbers in, 111, 119, 126; marketing cattle in, 131; settlement of, 16
Piedmont Substation, 123
Pike County, 21, 36, 85, 118, 181 (n. 74)
Piney woods region, 2, 7, 48, 95, 123, 135, 164; cattle round-ups in, 19, 111; cotton planting in, 16, 26, 27; decline of cattle numbers in, 134; demise of open-range herding in, 134; depletion of cattle during Civil War in, 37, 41; increase of cattle numbers in, 118; open-range herding in, 10, 11, 12, 15, 17, 18–19, 20, 25–26, 27, 31, 45, 46, 111–12; settlement of, 16; tick eradication in, 61–63
Plantation Saddle Horse Association of America, 120
Planters Cotton and Cattle Credit Corporation, 120
Poillnitz, Charles A., 48
Polk, Leonidas, 35
Poll tax, 182 (n. 21)
Pollard, Ala., 36
Polled Hereford cattle, 89, 107, 148, 149, 150, 153, 154. *See also* Hereford cattle
Pope, John, 9
Pork: Confederate ration of, 181 (n. 72); decline of availability during Civil War, 30; preference for in South, 21, 27, 48; supply of in Confederate commissary, 41
Poultry, 6, 20, 90, 109, 110, 125, 128, 153, 158, 164
Powell, William, III, 164, 170
Prattville, Ala., 66
Prices. *See* Cattle prices
Production credit system, 96–97, 99
Progressive Farmer, 103
Prunty, Merle, Jr., 90, 139
Purebred cattle. *See* Thoroughbred cattle

Quarter horses, 135
Quisenberry, Ralph D., 72

Railroads: role in expansion of cotton planting, 48–49, 56; role in closing the range, 51, 54
Randall, H. P., 186 (n. 73)
Randolph, Walter, 121
Randolph County, 21, 26, 49, 118, 194 (n. 53)
Randon, John, 10
Reconstruction Finance Corporation, 97
Red Hat Feeds, 146
Red Poll cattle, 78
Regional Agricultural Credit Corporation, 97
Reid, Tom C., 121, 135
Rifkin, Jeremy, 163
Riley, Benjamin Franklin, 17
Rittenour, Charles W., 69, 96
Robertson, G. W. "Billy," 140, 170
Rodeos, 135

INDEX

Romans, Bernard, 8
Ross, Isaac, 56
Roswald, Simon, 72
Rousseau, Lovell H., 33, 37
Rucker, John, 154
Ruffin, Edmund, 24
Rural Electrification Administration, 122
Russell County, 17
Rustling, 36–37, 106, 134, 135
Ryall, J. S., 47, 53
Rye grass, 130

S & C Beef Processors, 164
St. Clair, R. J., 109
St. Clair County, 94, 156, 181 (n. 74)
St. Louis, Missouri, 80, 95
St. Stephens, Ala., 27, 181 (n. 73)
Saloutos, Theodore, 101
Salting, 7, 9; railroad tracks, 54
Samford University, 140
Sand Mountain Livestock Market, 155
Sand Mountain Reporter, 154
Sand Mountain Substation, 90
Santa Gertrudis cattle, 140, 149
Santo Domingo, 3
Scottsboro, Ala., 155
Screw worm, 99–100, 127, 147
Seddon, James, 39–40
Selma, Ala., 30, 37, 38, 68, 83, 92, 106, 108, 116, 117, 149
Selma Creamery and Ice Company, 83
Selma Stockyard, 92, 131
Sericea lespedeza, 108, 130
Sexton, Alonzo, 46
Shedd, H. P., 88
Sheffield, Selden, 117, 140, 170
Shelby County, 154
Shoals Cheese Company, 124
Shore, Dinah, 163
Shorthorn Association, 69, 74
Shorthorn cattle, 23, 24, 25, 69, 71, 72, 73, 78, 79, 94, 106, 116, 119, 147, 149, 186 (n. 66), 186 (n. 74)
Shuptrine, Cecil, 149
Silliman, E. R., 74
Simmental cattle, 149
Simmons, A.D., 186 (n. 73)
Sims, W. Comer, 170
Slaves as cow hands, 9, 22
Sloan, William, 21

Smart, Franklin, 148
Smith, Dr. George C., 171
Smith, Lawrence, 46
Smith, M. F., 186 (n. 73)
Smith, McQueen, 66
Smith, Thomas G., 27
Smith, Will Howard, 66–67, 79, 170
Smith, W. S., 34
Smith-Lever Act, 64
Snow, Harry E., 81, 89, 91
Soil conservation, 92, 96, 100, 102, 103, 108, 109. *See also* Terracing
Soil Conservation and Domestic Allotment Act, 102
Soil Conservation Service, 100, 101, 108, 116, 118, 154
Soil Erosion Service. *See* Soil Conservation Service
South Carolina, 4, 15
Southeastern Livestock Association, 97
Southeastern Livestock Exposition, 136
Southeastern World Championship Rodeo and Livestock Week, 136
Southern Cattle Company, 80
Southern Cattlemen's Club, 60
Southern Cultivator, 39, 40
Southern Dairies, 123
Southern Plantation, 188 (n. 18)
Southern Railway, 110, 147
Soybeans, 108, 109
Spanish: cattle breeds, 10, 58, 79, 135, 148; in Florida, 1–2
Spurlin, Frank, 117
Stallworth, M. C., Jr., 133, 141, 148, 170
Starkville, Mississippi, 58
Stedham, Moses, 12
Steele, A. F., 47
Stevenson, Ala., 33
Stickney, Hazel, 66 71, 86, 88, 92, 138, 139
Stock laws. *See* Open range herding: demise of
Stockyards and Brands Division (State Department of Agriculture), 134
Stokely, J. T., 95
Stoner, C. I., 69
Strachan, Patrick, 4
Stuart, James, 9, 16
Stuart, John, 5
Subsistence Bureau, 39, 41
Sugg, R. S., 89, 99, 104, 107, 121
Sulphur Springs, Ala., 63

Summers, A. D., 71
Sumter County, 53–54, 58, 70, 71, 138, 139, 156, 157, 168, 184 (n. 41), 185 (n. 57)
Surplus Commodity Corporation, 99
Suttle, J. Freeman, 82
Suttles family, 92
Suttles plantation, 139
Swan, Samuel, 23
Swift and Company, 81, 99
Switch cane. *See* Canebrakes
Swoope, W. C., 73
Sylvan Agricultural District, 50

Tait, James A., 21, 22
Talladega, Ala., 106
Talladega County, 49, 181 (n. 74)
Tallapoosa County, 101
Tallapoosa River, 6, 7
Tallassee, Ala., 6
Tallow, 12, 28, 128
Tartt, T. M., 71
Tatum-Embry Company, 80
Tax-in-kind, 32, 36, 40–41
Teague, William H., 81
Tennessee, 35, 80, 93, 94
Tennessee Coal, Iron and Railroad Company, 116, 119–20, 121
Tennessee River, 35
Tennessee Stove Works, 155
Tennessee Valley, 7, 26, 28, 30, 37, 48, 80, 110, 11, 120, 151; agricultural research in, 58; antebellum breed improvement in, 21, 22, 23, 24; cattle auctions in, 131; Civil War in, 32–35; cotton planting in, 9, 20, 29, 76, 109, 157; depletion of cattle in, 41; drovers in, 47; Grade A dairies in, 123; growth of Alabama Cattlemen's Association in, 137; herders in, 15, 20; increase of cattle numbers in, 111, 118, 119, 126, 137, 156; planter-livestock raisers in, 22, 29, 76, 109, 141; settlement of, 16; thoroughbred cattle in, 58, 73, 79, 109; tick eradication in, 59, 60
Tennessee Valley Authority, 100, 101, 118, 122
Tennessee Valley Substation, 90, 122
Tensaw River, 4, 8, 10
Terminal markets, 80, 131. *See also* Marketing cattle
Terracing, 100, 101, 108. *See also* Soil conservation
Texas, 3, 14, 27–28, 32, 38, 42, 58, 64, 71, 78, 90, 98, 99, 140, 145, 149

Texas fever tick. *See* Cattle tick
Thomas, Carl B., 133, 170
Thoroughbred cattle, 22, 23, 57–58, 66, 72–73, 79, 89, 105, 106, 109, 116, 117, 127, 186 (n. 75)
Tichenor, Isaac, 55
Tombigbee River, 3, 8, 12, 17
Tondera, Steve, 171
Tonsmiere, Arthur, Jr., 133, 141, 170
Toulmin, Harry, 11–12
Tri-States Fat Stock Show, 106
Trotman, John M., 140, 170
Trouillets (free black family), 12
Trowbridge, J. T., 30
Troy, Ala., 60
Tubbs, Elbert, 52
Tubbs, R. S., 186 (n. 73)
Tupelo, Mississippi, 123
Turner, Frederick Jackson, 14
Tuscaloosa, Ala., 37, 125, 194 (n. 53)
Tuscaloosa County, 50, 83
Tuscumbia, Ala., 73
Tuskegee, Ala., 107
Tuskegee University, 64, 183 (n. 33)

Union Springs, Ala., 106, 116, 117, 148
Union Stock Yards, 80, 81, 89, 92, 93, 95, 105, 106, 110, 131
Uniontown, Ala., 56, 57, 58, 65, 131, 140
United States Commissioner of Agriculture, 41
United States Department of Agriculture, 58, 59, 64, 101
United States Department of the Interior, 101
University of Alabama, 67, 68
University of Illinois, 104, 120
University of the South, 67
USA Today, 163

Vandegrift, E. N., 153, 154
Vetch, 71, 101, 108
Veterinarians, 127; College of Veterinary Medicine, Auburn University, 59

Wadsworth, Edward, 107–8, 139, 170
Wadsworth Brothers Farm, 139
Walk, Charlie, 47
Walker, J. J., 39
Walker County, 52, 61, 119, 154, 156, 157, 194 (n. 53)
Wallace, George, 150
Wallace, Henry A., 99

INDEX

Wallace family, 69, 140
War of 1812, 12
Warren, Henry, 46
Warrior Agricultural District, 50
Washington County, 10, 18, 26, 27, 37, 45, 94, 111, 118, 133, 134, 181 (n. 73)
Watkins, Carrithers, 80
Watson, Fred S., 17
Watts, Thomas H., 36, 37, 39–40
Waugh, W. V., 186 (n. 73)
Webb, C. A., Jr., 108
Webb, J. C., 108
Webb Brothers Plantation, 108–9
Weiss Implement Company, 161
Wendland, Milton, 170
Wentworth, Edward, 117
West Alabama Fair, 24, 49
West Dallas Farms, 67, 69
West Indies, 19
Wetumpka, Ala., 6
White, L. H., 80
White clover, 91, 103, 109, 130, 154, 159
White-faced cattle. *See* Hereford cattle; Polled Hereford cattle
Whitfield, B. W., 48
Whitfield, Gaius, Sr., 22, 47, 48
Wilcox County, 21, 67, 94, 120, 137, 140, 146, 157, 168, 181 (n. 74), 184 (n. 41), 185 (n. 57), 194 (n. 53)

Wilson, Edward Hamilton, 132, 133, 136, 147, 150, 163, 164
Wilson, James H., 37–38
Winchester, Tennessee, 94
Winston, William, 53
Winston County, 46, 49, 110, 156, 194 (n. 53)
Wiregrass: cattle auctions in, 131; cotton raising in, 49; country buyers in, 193 (n. 41); dairy industry in, 125; decline of cotton raising in, 152–53; growth of Alabama Cattlemen's Association in, 137; hog farming in, 64–65; increase of cattle numbers in, 111, 118–19, 126, 152–53; open-range herding in, 18; part-time farming in, 164; peanut farming in, 64–65
Wiregrass Substation, 90
World War I: impact on cattle industry, 61, 76, 77–84, 86
World War II: impact on cattle industry, 115–19, 122–23

Yellow hop. *See* Hop clover
Yeoman farmers, 10, 12, 14, 15, 25, 26, 28, 29, 47
York, Ala., 72
Young, Roland, 103

Zeigler Packing Company, 117
Zeigler v. South & North Alabama Railroad Company, 54